GWYNNE DYER has worked as a journalist, broadcaster and lecturer on international affairs for several decades. He writes a twice-weekly syndicated column on international affairs, published in a hundred newspapers in forty-five countries and translated into more than a dozen languages.

www.gwynnedyer.com

INTERVENTION
EARTH

ALSO BY GWYNNE DYER

Climate Wars
After Iraq
War
Growing Pains: The Future of
Democracy (and Work)
Don't Panic! Isis, Terror and the Making
of the New Middle East
The Shortest History of War

Published in Great Britain in 2024 by
Old Street Publishing Ltd
Notaries House, Exeter EX1 1AJ

www.oldstreetpublishing.co.uk

ISBN 978-1-91308-326-7
Ebook ISBN 978-1-91308-327-4

10 9 8 7 6 5 4 3 2 1

A CIP catalogue record for this title is available from
the British Library.

Printed and bound in Great Britain.

INTERVENTION EARTH

EARTH

LIFE-SAVING IDEAS FROM THE
WORLD'S CLIMATE ENGINEERS

GWYNNE DYER

CONTENTS

Note on the Text

For commercial reasons, Tina Viljoen's name does not appear at the front of this book, but from shooting every interview to debating all the available options for coping with the crisis, she has had an equal share in making it happen. Working with her on this and many other projects has been the delight of my life.

All direct quotations, unless otherwise noted, are taken from interviews with the more than 100 climate scientists, engineers and a few others who generously gave their time. There is a list of all the interviewees at the back of this book.

INTRODUCTION

In 2008, I wrote a book called *Climate Wars* about the science and the geopolitics of climate change, and for a couple of years afterwards I had a sort of intermittent double vision. In my mind's eye, I would suddenly see the three-degree-hotter world overlaid on the world that existed at that time. Such afflictions, however, can usually be cured by a judicious combination of outdoor exercise and good wine – and now that you know the remedy you can be confident that reading this book will cause no similar derangement. But when I immersed myself in that world again more than a decade later, I did notice that some of the scientists I'd met first time around now seemed a bit more – what's the word? – distracted. Not actual waking visions, you understand. Just moments lost in thought.

As the climate crisis deepens and the negative impacts multiply, public opinion and politics are finally responding, but there is no guarantee that our actions will be big and fast enough to avoid an outcome that is catastrophic for human beings and quite disruptive, at least, for the entire biosphere. We are not even sure yet how big and how fast those actions need to be, because the discipline of climate science is only about forty years old. But the answer is almost certainly: very big and very fast.

How extreme it could get depends on two main things: how intent we are on burning all the fossil fuels we already know about, and how sensitive the Earth's climate is to that carbon injection. We could get to 8°c of warming fairly readily, but we probably wouldn't, because it would be so catastrophic well before we reached that point that it would terminate our activities. We are still able to trigger warming of the order of 5°c globally, burning only a fraction of the fossil fuels, if we consider the feedbacks and tipping points. Very sobering for those of us who work on this day in, day out.

Tim Lenton, Professor of Climate Change and Earth
System Science, University of Exeter

Scientists can now predict with some confidence how much extra carbon dioxide in the atmosphere will cause how much warming: there is still a range of possibilities, but the range has narrowed and all the possibilities past 450 parts per million (ppm) of carbon dioxide (CO_2) in the atmosphere are bad.[*]

We are coming up on 425ppm, and adding 2.4ppm a year. Scientists can even foresee how fast that warming will happen – *if they can assume that this will always be a steady, 'linear' process.* But we now know that warming is often 'non-linear': that is to say, the average global temperature crosses an invisible threshold, a sort of tripwire, and makes a sudden unscheduled leap upwards. 'Tipping points' are the specific points at which the upward leaps occur, but climate scientists have only a vague and uncertain knowledge of where they are. Two major questions addressed by this book, therefore, are the speed and accuracy with which scientists can map the 'tipping points', and how we might find ways to avoid crossing them even now.

Since my last foray into the field, there has been a significant loss of faith in the notion that emission cuts alone can stop us short of reaching the tipping points. There is an emerging

---

[*] References to the amount of carbon dioxide in the atmosphere in this book will, unless otherwise stated, mean 'carbon dioxide equivalent'. That is, the sum of all the 'greenhouse gases' in the air that cause warming, including methane, nitrous oxide, etc., expressed in terms of the warming that would be caused by an equivalent amount of carbon dioxide alone.

debate – emerging into public view, at least; it has been raging inside the climate science community for some time – about which methods of direct human intervention into the workings of the climate system would be valid and safe, and which would not – in other words, about 'geoengineering' or 'climate engineering'. (The former term is generally used in North America; the latter is preferred in Europe and elsewhere.)

This debate has become so fraught that quite a few climate scientists who favour only 'Carbon Dioxide Removal' (CDR) techniques such as Direct Air Capture (DAC) or Bio-Energy with Carbon Capture and Storage (BECCS) now want to remove these technologies entirely from the category of 'geoengineering' techniques with which they have usually been grouped. This would be done to distinguish the CDR technologies more sharply from the allegedly more dangerous but generally cheaper and quicker Solar Radiation Management (SRM) techniques that involve direct human intervention to reduce the amount of solar energy reaching the planet's surface. There are more and less desirable techniques within each category, but this is increasingly where the battle lines are being drawn: between CDR on the one hand, and more direct interventions like SRM on the other.

Almost nobody in the climate science community really believes any more that we can stop the warming at a place that is relatively safe without direct human intervention of some sort in the climate system. Doing so merely by cutting emissions and planting lots of trees would have been possible (with a huge crash programme) in the year 2000, and it was still imaginable (just) in 2010, but it now hardly seems credible.

This book will assess the belief that some form of intervention in the climate system (CDR, SRM or a combination of the two) will be required to stop the warming short of a catastrophe.

A 'never-exceed' target that most governments now agree on, high though it may be, is an average global temperature 'less than two degrees Celsius higher than it was in pre-industrial times'. (The shorthand for this is <+2°C.) An 'aspirational' target of only <+1.5°C was adopted by the Intergovernmental Panel on Climate Change in 2018, but that is already teetering on the edge of impossibility. If we overshoot +2°C, we are likely to enter a chaotic world in which sudden upward lurches in temperature are added to the relentless current rise (officially estimated at +0.18°c per decade), and the hurricanes, forest fires, killer heatwaves and the rest will grow correspondingly more severe and more frequent.

> The 1.5°C target is one that science increasingly demonstrates is associated with substantial risk of triggering irreversible large change and that crossing tipping points cannot be excluded even at lower temperature increases.
>
> Katherine Richardson, Director, Sustainability Science
> Centre, University of Copenhagen

In other words, we are already in the danger zone.

In recent years, the various methods for reducing emissions have improved greatly both in affordability and in variety. To take just two examples: solar power has become dramatically cheaper, while meat substitutes and 'cultivated' (lab-grown) meat are being developed which would, in theory, enable us to rewild huge amounts of pastureland now devoted to feeding beef cattle, to the great benefit of both biodiversity and emissions cuts. But in every case, the question that must be answered is: 'When will this solution be available at scale?' Because the 'never-exceed' deadline is drawing near.

This book is not a consciousness-raising exercise, and there will be no exhortations to pull up our collective socks and get on with 'saving the planet'. Nor will I waste people's time with a lengthy trudge across the now familiar territory of the climate emergency; others have already done that. But just as an *aide-mémoire*, here are the inconvenient facts that we must always bear in mind.

FACT NO. 1: We Are Running Out of Time

Actually, we probably *have* run out of time. Like soon-to-be-bankrupts, we can go on fiddling the books for a while longer, but we cannot stay below the 1.5°C higher average global temperature that was our recommended maximum increase according to the Paris Climate Agreement of 2015. As Johan Rockström, director of the Potsdam Institute for Climate Impact Research, told me in 2020:

> We have been lulling ourselves into a comfort zone, believing we have a lot of time, but 2020 is the year when we need to bend the curve down on global emissions, because when you look at the more than 100 scenarios in the Intergovernmental Panel on Climate Change (IPCC) report – the 100 different scenarios that could [stop the warming at] +1.5°C – they all bend in 2020. You cannot succeed if you bend later. Everything is determined by the carbon budget... If you bend later, the speed by which we have to reduce emissions is no longer possible to achieve in any democratic way. You would simply have to bulldoze every coal-fired plant overnight.

Well, the emissions curve did *not* bend down in 2020, despite the Covid-19 pandemic, and they haven't started bulldozing coal-fired power plants either. Global carbon dioxide emissions did drop briefly – by 17 per cent – at the peak of the first wave of Covid-19, but over the whole year the needle barely flickered. The planes stopped flying for a while, but the cows kept burping, the lights stayed on, and the houses of the developed world remained warm in winter and cool in summer.

For all the talk of cuts, the amount of carbon dioxide in the air has increased almost every year since the start of the industrial revolution. In 1800, it was only 280 parts per million (ppm). By 1988, when global warming first became a public concern, it was 350ppm. In 2020 it was 415ppm, and it's still going up. There is little chance that the curve will turn down before 2025 at the earliest – whereas achieving the 'aspirational' target of not exceeding +1.5°C would have required an already implausible cut of 7.9 per cent in greenhouse gas emissions in each year of this decade, starting in 2021.

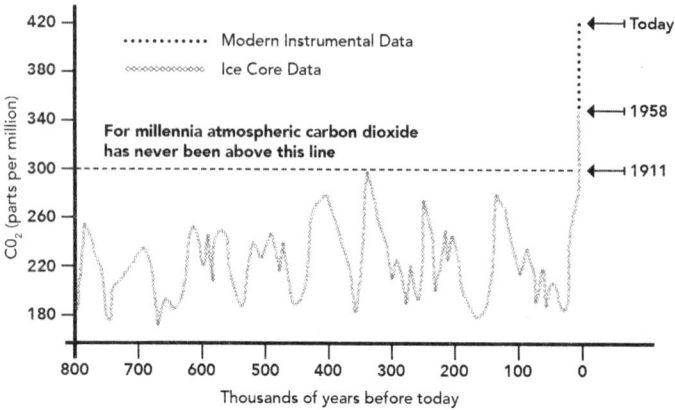

FACT NO. 2: Cutting Emissions Is Not Enough

There's a dirty secret about the Paris deal and the <+1.5°C 'aspirational' limit: the target could never have been achieved by cutting emissions alone. It is abundantly clear from many sources that the negotiators in Paris were counting on 'negative emissions' technologies – that is, taking greenhouse gases out of the air – to avoid some of the warming.[1]

This is a problem because almost all of these 'Carbon Dioxide Removal' (CDR) technologies are either slower-acting or much more expensive (or both) than simply cutting CO_2 emissions, and most are not yet ready for deployment at a global scale. Half of them also have major implications for land use or the health of the oceans. Some of these CDR technologies do have longer-term possibilities as part of an attempt to stabilise the global climate, and I will discuss them in detail later, but they cannot be deployed fast enough to help us stay below the <+1.5°C target into the mid-2030s.

FACT NO. 3: Carbon Accumulates

So much for the aspirational goal. How about the real, 'never-exceed' goal of stopping the warming at or before +2°C?

The carbon dioxide that we put in the air stays there for a very long time: 200 years for the average CO_2 molecule. Plants absorb some CO_2 each spring and summer as they grow, but they put it back into the air again when they die and burn or rot. Even the rocks absorb some CO_2, very slowly – but these natural 'carbon sinks' are largely occupied with playing their role in the natural 'carbon cycle'. Much of the CO_2 that human beings put into the air each year stays in the atmosphere and accumulates: even some of the CO_2 emitted by the coal-burning boilers on Thomas Newcomen's eighteenth-century steam pumps is still there.

Now, 450ppm of CO_2 in the atmosphere is the point at which we are effectively committed to +2°C. Beyond that, very bad things begin to happen. With the amount of CO_2 in the air already at 425ppm, we only have 25ppm left before +2°C average global temperature becomes inevitable. The extra amount of CO_2 emissions caused by human activities that accumulated in the atmosphere in 2022 was 2.4ppm. If we continue at that rate, we will reach 450ppm around 2032. Even if we cut our emissions by half in the next decade – a heroic but unlikely achievement – we would still reach at least 435ppm by the middle of the decade (2035).

Nobody in their right mind would willingly go to 435ppm, because there is not always a predictable, direct relationship between parts-per-million of CO_2 and global average temperature. At various points as the planet warms – unfortunately we don't know precisely which – 'tipping points' will be triggered and the global average temperature will lurch rapidly upwards. Most climate scientists – and the IPCC's official 'best guess' – assume that these thresholds are almost all higher than +2°C/450ppm, and it might be that the climate really spins out only at +2.2°C. On the other hand, the true 'never-exceed point' could also easily be +1.8°C, in which case 435ppm would be more than enough to cook our goose.

Yet these figures feel so small that it's hard to take them seriously. What's the difference between 1.8 and 2.2? Or between 435ppm and 450ppm? Well, it's similar in effect to the difference between a human body temperature of 36.5°C (normal), 38.5°C (fever), 40.5°C (brain damage), and 43°C (death). So yes, take it seriously. The better informed people are, the more frightened they are.

FACT NO. 4: Predicting the Climate is Hard

As meteorologist Edward Lorenz realised in 1960, if a butterfly flaps its wings in a certain way in Beijing in March, then by August hurricane patterns in the Atlantic could be completely different. The climate system is so complex and so interconnected that we cannot predict the weather for even one week, so how can we possibly predict the climate? Alan Robock,

Distinguished Professor in the Department of Environmental Sciences at Rutgers University, New Jersey, explains:

> We don't have any data on the future, and there's a lot of chaos in the climate system. We can predict the 'envelope' of possible weather, but not the specific weather. Then there's natural variability: some years are warmer than average; some are colder. Some years you get El Niño, some you get La Niña, and you can't predict those very far in advance. This is a problem that climate scientists have always had. We don't have a laboratory with test tubes and accelerators to do our experiments; the laboratory is the real world, and the best we can do is the climate models we create.
>
> We write down the equations that describe everything we understand and do multiple runs with slightly different initial conditions, putting in the flapping of butterfly wings and so on, and we get a swarm of potential climates. The real world will only go through one of those potential climates, but probably it will be somewhere within that swarm. Then we test those models on the past. If they do a good job simulating the effects of known volcanic eruptions, or if they do a good job simulating the global warming of the past century, then we have more faith in them for the future.
>
> Alan Robock

That's all we have, so it will have to be good enough.

FACT NO. 5: Averages Lie

The 'Average Global Temperature' is an indispensable concept when discussing the broad topic of 'global warming', but it is very unreliable as a guide to what the temperature will be in any specific location. Moreover, there is a big difference between temperatures at sea and on land. Temperatures are generally more extreme on land, because it heats up more quickly in sunshine and loses heat more quickly at night and in winter. The further away from the sea, the truer this is, which is why it's deep in the interiors of the continents that most of the record temperatures, both high and low, have been observed.

But since two-thirds of the planet's surface is covered by oceans, the Average Global Temperature is always closer to the average temperature over the oceans than it is to the average land temperature. These values are not usually calculated, but a rise in average global temperature of 2.0°C really means a rise of roughly +1.0°C in average *maritime* temperature and a rise in average *land* temperature of between +3.0°C and +4.0°C (depending mainly on how far inland).

FACT NO. 6: The Atmosphere Does Not 'Bounce Back'
Even if we do manage to achieve Net Zero by 2050, that doesn't mean everything goes back to 'normal'. We would have stopped adding more greenhouse gases to the atmosphere each year, but all the carbon dioxide that drove the temperature up to +2.0°C or more would still be there, and it won't leave of its own accord. If we want our old climate back and we are not willing to wait thousands of years for the rocks to do the job, we'll have to take the excess CO_2 out of the air ourselves – a massive, centuries-long task.

Since the alternative is living indefinitely with the brutal climate of a +2.0°C world, we will probably try to do that. Indeed, that is likely to be the long-term role of the various 'Carbon Dioxide Removal' (CDR) techniques now being researched or, in a couple of cases, developed.

FACT NO. 7: 'Runaway' Is Possible

Terms like 'runaway' and 'hothouse Earth' do not mean Venus-like conditions inhospitable to all life. Our planet is considerably further from the sun than Venus is. It will not experience the extreme conditions of that planet until the sun has heated up another 6 per cent, around a billion years from now. But if tipping points cascade, a rise in average global temperature of +4°C or more is possible by the end of this century. Low-probability but high-impact events are precisely what you buy insurance for, but unfortunately they tend to be omitted from most official climate documents.

> Human beings are a tough species. A few breeding pairs are bound to survive.
>
> James Lovelock

Temperature rises of up to +6°C would still not mean the extinction of the human race throughout its range (pretty much the entire land surface of this planet), but it would drastically shrink the climate niches where humans could survive, implying a die-back in global population of perhaps 90 per cent. Hundreds of thousands or even millions of other species would become extinct, but such temperatures and mass extinctions have happened before and this would not be The End. Our current civilisation would be unlikely to survive, and it might become impossible to build another one, but actual human extinction is quite unlikely.

This future is not yet inevitable. A hyper-aggressive worldwide programme of emissions cuts *combined* with the super-charged development and deployment of CDR techniques capable of extracting vast amounts of CO_2 from the air and getting rid of it somehow, *might* make it possible to stay below +2°C even into the 2040s – and by then, like stepping stones to the future, better means for reducing emissions and removing carbon dioxide from the atmosphere might have become available.

If that doesn't happen, the same goal of staying below +2°C might be achieved quite quickly, even at the next-to-last moment, by reflecting back enough incoming sunlight (aka Solar Radiation Management or SRM) to cool the planet's surface by a degree or two. This could not be a permanent solution, but it might win us a few extra decades to work on reducing emissions and deploying CDR techniques without crossing the tipping points and without suffering extreme warming that would topple global civilisation into famine, mass migration and war.

* * *

We probably came very close to +1.5°C in 2023 and may exceed it in 2024, but this is largely because an El Niño event came along and piled additional warming, unrelated to climate change, on top of the existing climate trend. (El Niños occur every three to seven years and the previous one was in 2016. That, too, was at the time the hottest year ever.) Many people will be hurt by the extreme weather of this period, and almost everybody will experience some discomfort – floods, droughts, forest fires, heat-waves, massive storms, etc. – but this is not climate Armageddon; it is a transient event. The average global temperature ought to drop back down to +1.3°–1.4°C by 2025, and the current assumption is that it will not reach and remain above 1.5°C until 2029.

* * *

The first part of the book, *The Planetary Trajectory*, explores the recent research into non-linear effects that has forced scientists to conclude that 'runaway' warming is indeed possible, has happened in the past, and could be triggered by far less human-caused heating than we had previously thought. The early chapters will therefore focus on the science that imposes this new urgency, on the fossil fuel lobby's ongoing attempts to discredit or distract from the science, and on the scientists who have expanded our understanding of the climate system and the entire Earth System.

Next, in *All About Emissions*, I will examine our efforts to end greenhouse gas emissions (with a focus on carbon dioxide and methane), and also the new technologies for 'negative emissions' (taking those gases back out of the air). The aim is to arrive at a realistic estimate of how much emissions can be cut and how quickly – and this requires a case-by-case examination of how much each technique, from Artificial Ocean Alkalinisation to growing the algae *Asparagopsis armata*, can contribute to the effort. Will our present emissions reduction strategy work, or will more radical measures will be required?

The third part, *Climate Geopolitics*, is about how the political environment will evolve as the climate worsens. Politics is the way mass societies must make their decisions, and that will not change even in the most extreme emergency. There are plenty of good ideas around – to stop glaciers sliding into the ocean, to expand the food supply, to suck carbon dioxide out of the air and bury it – but they will have to be prioritised, negotiated and paid for by groups of frightened, desperate people. No matter what we do now, the climate will go on getting worse for at least another twenty years, and human beings don't typically respond well to this sort of challenge. We will have to learn to act like grown-ups all the time.

By the final part, *Desperate Scientists and Last-Ditch Ideas*, I hope to have shown that we are likely to run out of time faster than many of these promising ideas can be deployed at scale to save us. At some point very soon, I will argue, we must have the great SRM debate. Indeed, we should start right now, because our whole future may depend on it.

PART ONE

THE PLANETARY TRAJECTORY

... in which it is revealed that the Earth System is less stable than was formerly supposed, and that the risk of 'runaway' warming is real

1

'HOTHOUSE EARTH'

Once upon a time all we had to worry about was the greenhouse gases relentlessly building up in the atmosphere, slowly but surely raising the average global temperature. Those were the Good Old Days, before the alarm was sounded by a very influential study published in 2018. You can access it directly on the web, but as it was written by scientists for scientists, the language can be a bit challenging.[2]

> We explore the risk that self-reinforcing feedbacks could push the Earth System towards a planetary threshold that, if crossed, could prevent stabilisation of the climate at intermediate temperature rises and cause continued warming on a 'Hothouse Earth' pathway even as human emissions are reduced. Crossing the threshold would lead to a much higher global average temperature than any interglacial in the past 1.2 million years and to sea levels significantly higher than at any time in the Holocene. We examine the evidence that such a threshold might exist and where it might be. If the threshold is crossed, the resulting trajectory would likely cause serious disruptions to ecosystems, society and economies. Collective human action is required to steer the Earth system away from a potential threshold and stabilise it in a habitable interglacial-like state. Such action entails stewardship of the entire Earth System...
>
> From 'Trajectories of the Earth System in the Anthropocene', published in the *Proceedings of the National Academy of Sciences of the United States*[3]

That abstract set off a cascade of alarm bells, not only in my head but in the heads of thousands of other people around the world. Let me translate some of the passages that set the bells ringing into non-scientific English.

'self-reinforcing feedbacks'
Natural and unstoppable sources of further warming triggered by human-caused heating.

'prevent stabilisation… at intermediate temperature rises'
We're heading towards the +2°C limit, and the brakes have failed.

'"Hothouse Earth" pathway'
Runaway warming.

'continued warming… even as emissions are reduced…'
We have lost control.

'much higher global average temperature'
+4°C, +5°C, +6°C, and on up.

'sea levels significantly higher'
Up to two metres sea-level rise in this century (goodbye Miami, Lagos, Calcutta, Bangkok); up to 5 metres by 2200 (goodbye New York, London, Mumbai, Tokyo).

'serious disruptions to ecosystems, society and economies'
Famine, war, mass refugee flows, mass die-backs of human and animal populations.

'stabilise it in a habitable, interglacial-like state'
We can't go back to the old climate, but we might be able to stop in one that is still habitable.

'stewardship of the entire Earth System'
We must become Planetary Maintenance Engineers.

The phrase 'planetary maintenance engineers' popped into my head as soon as I read the abstract. It comes from James

Lovelock's first book about the 'Gaia hypothesis' – now known in universities as 'Earth System Science' – published in 1979. At that time, Lovelock wasn't talking specifically about climate, because global warming wasn't yet high on the scientific agenda. He was discussing how Gaia (the self-regulating 'Earth System' that he identified and named) was being slowly crippled by the ever-increasing pressure of human activities:

> If man had encroached upon Gaia's functional powers to such an extent that he disabled her... he would then wake up one day to find that he had the permanent lifetime job of planetary maintenance engineer... The ceaseless, intricate task of keeping all the global cycles in balance would be ours... and whatever tamed and domesticated biosphere remained would indeed be our 'life support system'.

This was the first time any serious scientist publicly suggested that human beings might have to step in and manage the planet's many autonomous and hitherto self-regulating biogeochemical systems. Lovelock wasn't suggesting it with any pleasure, because forty years ago we lacked the knowledge, tools and sheer processing capacity to do that job (as we still do today). But we are nearing the point when we may have to take on at least some of that ghastly, thankless job, ready or not, for we are overloading and disabling the systems that kept the climate and much else stable. It is much less uncommon today than it was in 1979 to hear scientists talk, in some desperation, about our having to take on 'the stewardship of the entire Earth System'. So who wrote this remarkable 'Trajectories' paper, and why?

There were sixteen authors, all leading climate scientists working in eight countries.[4] The paper contained no original research, no new facts. All it did was summarise and analyse the recent research in the field (admittedly a monumental task) and draw the obvious conclusions. Yet it had a huge impact, with a sky-high 6,061 score on Altmetric, which tracks the impact of academic articles. This kind of massive, instantaneous response only occurs when you say in public for the first time what a great many people are already uneasy about in private.[5]

'I knew the paper would just explode in the faces of people interested in climate science,' said Hans Joachim Schellnhuber, Director Emeritus at the Potsdam Institute for Climate Impact Studies. 'Persuading somebody of a completely different world-view is impossible, but to bring out some of the things they already know deep inside – or not so deep inside – that's possible.' This phenomenon is common in fields like climate change, where the conclusions of new research are often provisional and conditional. As the evidence accumulates, experts may begin to suspect or even fear that it adds up to a really big deal – but nobody wants to be the first to say so aloud. The media and the public are liable to grab the wrong end of the stick, and your own colleagues may conclude that you are just seeking attention. Scientific research is a highly competitive field, and scientists can be torn to shreds by their peers if they go one millimetre further than the evidence supports. Think piranhas.

By 2018, the evidence was piling up that the projections for future warming that the big climate conferences and treaties had been working with were much too optimistic, but in professional terms it would still have been unwise for an individual scientist to take all that evidence, drawn from many different domains of research, and offer sweeping conclusions about it. It was information that people didn't want to hear, and there would inevitably be a desire to shoot the messenger. What was needed was a major review paper by multiple authors, each an expert in some aspect of the accumulating research, and at least a few of them really big names in the discipline.

One such name was Tim Lenton, Professor of Climate Change and Earth System Science at the University of Exeter.

A fairly sizable group of us had been researching the possibility of different tipping points in the Earth's climate system, so it was natural that we would ask: 'Is there a global tipping point? Is there an instability of the whole climate?'

We burn a certain amount of fossil fuel, warm up the planet a certain amount, but then the feedbacks within the climate system start to amplify that until the climate change becomes almost self-propelling, because carbon is being released from degraded ecosystems and the permafrost and so on.

We weren't alone in being concerned, but we came together to raise a flag: this is a risk that can't be ruled out. And if it's a risk that can't be ruled out, let's have a go at trying to work out how big a risk it is.

                                                    Tim Lenton

Many of the 'Trajectories' authors had met in the 1990s as members of the scientific steering committee of the UN's International Geosphere Biosphere Programme (IGBP). Among them were Katherine Richardson, an American oceanographer now leading the Sustainability Science Centre at the University of Copenhagen, Hans Joachim Schellnhuber (German, founder of the Potstdam Institute), Paul Crutzen (Dutch, Max Planck Institute for Chemistry), who was awarded a Nobel Prize for discovering how the ozone layer was affected by human activities, Johan Rockström (Swedish, at the time studying systems ecology at Stockholm University), as well as the then director of the IGBP, Will Steffen (American, Australian National University).

In 2009, they had co-authored the paper 'Planetary Bound-aries: Exploring the Safe Operating Space for Humanity', which broached the notion of specific thresholds in various parts of the Earth System that, if crossed, could trigger abrupt and radical change. In the 2015 update, produced to coincide with the Paris climate summit, they reported that four of the nine planetary boundaries they had defined – climate change, bio-sphere integrity, biogeochemical flows and land system change – might already have been breached.[6]

Soon after the Paris summit, the group began working towards the key 'Earth Trajectories' paper of 2018. Only in retrospect did it become clear what had changed. Will Steffen explained:

> During the years leading up to [the 'Trajectories' paper], I was spending a fair bit of time in Stockholm working with colleagues at the Stockholm Resilience Centre, especially Johan Rockström [then the Director] and Katherine Richard-son from Copenhagen. We all had a common concern that despite a lot of research on feedbacks, on tipping points, in the biosphere, in the oceans and in the cryosphere, every time a big report came out from the physical climate guys it had very even trajectories completely dominated by human emissions, [as if] that was going to decide where your system went.
>
> At the Resilience Centre, they also do a lot of complex system thinking, so they're looking at states in transition, and it's not smooth changes, it's abrupt changes between alternate stable states.
>
> Will Steffen

Katherine Richardson gives a lot of the credit to Lovelock and his idea of Gaia: to systems thinking on the biggest scale, slowly but irresistibly reshaping the perspective of climate scientists (and of every other educated human being, whether they realise it or not):

> We have a geology department, a physics department, an economics department, a humanities department. The object is to find out as many details as you possibly can *in your box*. It's as if we believe that if you could put all of your details in a big pot together with everybody else's details from their boxes, then we'd understand how the Earth works...
>
> It's like asking doctors who have different specialties – heart and brains and lungs and reproduction and skin and feet – to put all their knowledge in a pot and stir it up. We still wouldn't know what a person is, because it's the interactions between them that are really important. And that's what James Lovelock did. He focussed on these interactions. He said that if you wanted to find life on another planet, you should just analyse the atmosphere around that planet, because there would be certain elements if there was life that wouldn't be there if it wasn't. He was one of the very early pioneer thinkers about systems. It's no accident that Earth System science didn't develop at a university.
>
> Katherine Richardson

When I spoke to him in 2021, James Lovelock agreed with Katherine Richardson:

> The source of the trouble here is, I'm afraid, the university system. It somehow made the ghastly mistake a long time ago that you cannot teach all the subjects together, you have to separate them, and you teach chemistry in one building, physics in another, engineering in yet another. The students

in each of these buildings never meet, and this produces an incomplete knowledge in the community at large. This is disastrous.

The final piece of the puzzle came in a book written by Marten Scheffer, a Dutch mathematician and expert on complex system theory. Scheffer had pointed out that during the glacial-interglacial cycle of the current Ice Age – 50–100,000-year periods of deep glaciation, followed by 10,000–15,000-year 'interglacial' periods like our own that were thought to be around 5°C warmer – the actual 'forcing' (that is, the changes in the amount of solar heating reaching Earth) is very small. The minor shifts in the Earth's orbit known as the Milankovitch cycles, which slightly alter the amount of energy being received from the sun, account for only about one degree C of the regularly repeated five-degree C change up or down in average global temperature. What had caught Will Steffen's eye in Scheffer's work was that the rest – '80–90 per cent of the heavy lifting' – was down to feedbacks within the Earth system, and they were non-linear. 'So why,' he asked, 'would we expect under pretty horrific forcing (like now) for the Earth to be well-behaved in a nice linear fashion?'

Steffen, Richardson and Rockström decided it was time to share their insights with the wider scientific community.

> We'd been mucking around with this for a while... publishing papers on tipping points. We didn't have a model yet, but we needed to put this out on the table, to say 'we think the Earth System could operate this way, not the way most people think.'

They flew a group of scientists to Stockholm for a workshop, including John Schellnhuber and Marten Scheffer. And it was on the plane that Scheffer came up with one of the critical figures in the Trajectories paper.

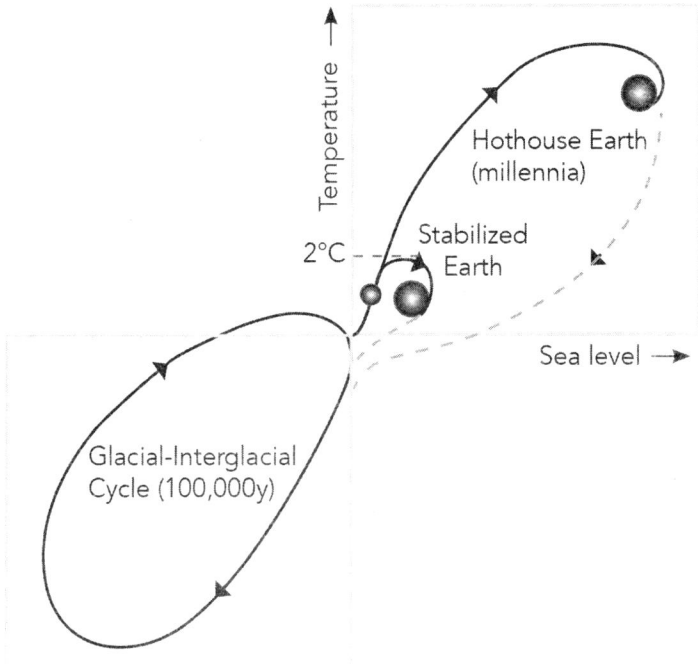

What he had in mind was a diagram showing 'limit cycles' of temperature, one loop describing the range of temperatures we live in now, with a transition, already underway, to a different limit cycle that is much hotter: 'Hothouse Earth'. Our planet might eventually return to its earlier, cooler limit cycle, but not for a couple of hundred thousand years – the time it would take for the carbonate and silicate cycles to remove the CO_2 from the atmosphere. Crucially, the diagram also showed a much smaller, tighter circle labelled 'Stabilised Earth' in which the planet's average temperature never exceeds $+2°C$. Not 'Stable', mind you; just 'Stabilised'.

Switching to the 'Stabilized Earth' pathway would require

'deep cuts in greenhouse gas emissions, protection and enhancement of biosphere carbon sinks, efforts to remove CO_2 from the atmosphere, *possible solar radiation management* [my italics], and adaptation to unavoidable impacts of the warming already occurring'.

The article, published in August of 2018, really did get the donkey's attention, both within and beyond the scientific world.

The Anthropocene

For years, climate scientists had been quietly panicking. Now the 'Trajectories' paper gave form and substance to their worst fears. It said two things that had not been said with such authority before. One was that we have already left behind us not only the 'Holocene' epoch, a period of stable, moderate climate, suitable for large-scale agriculture, that has prevailed since the last major glaciation ended 11,600 years ago, and in which we have built all of human civilisation. We have also left the whole Quaternary period behind: the alternation of long periods of glaciation with briefer 'interglacial' warm periods that has obtained for the past 2.6 million years. We are now on a new trajectory, heading up into the recently named 'Anthropocene' epoch, a hotter period when the pressure of human activity on the Earth System is the dominating factor in determining outcomes.[7]

The other even more dramatic conclusion is that human decisions that must be taken in the next couple of decades will determine how hot the Anthropocene is. Intervene intelligently and 'take charge' of the climate, and we could probably stabilise it at around 2°C higher average global temperature: a good deal hotter than now, and very hard on at least some people, but not utterly catastrophic. 'Stabilised Earth', as the authors call it.

But don't do enough, or do it too late, or do the wrong

things, and the average global temperature will end up as much as 5°C or 6°C higher ('Hothouse Earth'), with no way back on a time-scale relevant to human beings. As Johan Rockström explains:

> Thirty years of climate science has given us so much under-standing, and what I now see very clearly as a red thread during that entire journey is that the more we learn about the Earth system, the more reason for concern we have... In 2001, at the Third Assessment [of the IPCC], the best assessment of the risk of crossing catastrophic tipping points, of destabilising the biosphere, was estimated to be somewhere between +5°C and +6°C of warming. Even now, we believe we could avoid that. So you could argue that in 2001, twenty years ago, we assessed that the risk of catastrophic change and causing a tipping point was basically zero.
>
> Then for every new assessment the level [of average global temperature at which the risk of crossing tipping points gets serious] just goes down, down, down – until 2018, where the assessment is somewhere between +2°C and +3°C. People think we're raising the alarm because human pressures are increasing, but that's not the case at all. It's just that we are learning how the planet works, and the more we learn, the more we realise how vulnerable she is.
>
> When humans start this massive global experiment of putting pressure on the planet, with greenhouse gases and cutting forests and loading nutrients into the oceans, what does the Earth system do? It responds by buffering and dampening the impacts, just shoving our planetary debt under the carpet, because we are so far away from the tipping point that the systems have huge redundant capacity – what we call *resilience*. We've followed a 150-year journey since we

kicked off the industrial revolution and we've gradually been losing resilience, but up until recently the models have been right. Things change linearly when it's resilient, but when you lose resilience... Bang! Things can crack, and you tip over into new states.

That's what the 'Trajectories' paper is telling us: that promises of 'Net Zero by 2050' and other pie-in-the-sky-when-you-die schemes for eventually reducing emissions aren't going to cut it, because the crisis is coming a great deal sooner than that. It would be nice if all we had to worry about was the slow, steady, linear growth of emissions, but our bigger, much more urgent problem is figuring out how to avoid triggering the feedbacks in the next ten to twenty years.

Or to put it in the paper's scientific language:

In most analyses, [the Earth System's] trajectories are largely driven by the amount of greenhouse gases that human activities have already emitted and will continue to emit into the atmosphere over the rest of this century and beyond – *with a presumed quasi-linear relationship* [author's italics] between cumulative carbon dioxide emissions and global temperature rise...

But the 'quasi-linear relationship' is precisely what the 'Trajectory' authors say we should not presume:

Our analysis implies that even if the Paris Accord target of a 1.5°C to 2.0°C rise is met, *we cannot exclude the risk that a cascade of feedbacks could push the Earth System irreversibly on to a 'Hothouse Earth' pathway.*[8] [Author's italics]

Positive Feedbacks

A 'positive feedback' is what you get once you have crossed a tipping point: it's a self-reinforcing, self-amplifying process that human beings cannot stop. For example, as the Arctic region warms, some of the permafrost thaws. The permafrost is a permanently frozen layer of ground that lies beneath the 'active' surface layer that thaws and refreezes each year, and it contains large amounts of dead vegetable and animal matter that has been buried within it for centuries or millennia. If the permafrost layer starts to thaw at its southern edges (as it is doing now), it will release some CO_2. So long as the thawing is only happening on a small scale, however, those emissions would stop again if, in the near future, we stop dumping large amounts of greenhouse gases into the atmosphere.

But suppose that we go on heating the world with our emissions, the climate warms, and the permafrost thaws over a large area. The dormant bacteria in the soil revive and start consuming the dead organic matter, and as that rots it presumably releases the 'gases of decomposition' (carbon dioxide, methane, nitrogen and hydrogen sulfide) into the air in large quantities. Two of those, carbon dioxide and methane, are greenhouse gases, and they, too, begin to contribute to the warming. This is the feedback process in action: we didn't produce those gases directly, but it was the warming we caused by our own emissions that started the process rolling – and permafrost emissions could become truly massive.

There is a permafrost layer ranging from a metre to a kilometre thick beneath about a quarter of the land surface of the Northern Hemisphere, and it contains around twice as much CO_2 as there is now in the entire atmosphere. If we ever get that feedback moving fast, our own emissions will become almost irrelevant: we're heading for 1,500ppm, and we can't stop it.

Fortunately, this particular feedback is one of the slowest to activate on a grand scale, and it would take many centuries to thaw all the permafrost. But even a relatively small area of the permafrost thawing releases a very large amount of CO_2 and methane, and there are other feedbacks that act a lot more quickly. Especially since they may act in cascades.

Tipping Points

The tipping points are not a single row of dominoes (topple the first and all the rest go down), but they do come in clusters, and these clusters can cascade in a domino-like manner. The 'Trajectories' paper lists fifteen tipping points to worry about, such as die-back in the Amazon forest and in the boreal forests of the far North, permafrost thawing, increased bacterial respiration in the ocean, and the weakening of land and ocean physiological carbon sinks. Of course, there may be 'unknown unknowns' out there too.

By the time human emissions have caused 2°C of warming, that warming will itself have triggered feedbacks that add another 0.47°C to the average global temperature, for a total of (rounding up) +2.5°C – and maybe, by the year 2100, +3.8°C. There's some room to quibble on all these numbers, but mostly on the upside, and they clearly put us on the 'Hothouse Earth' trajectory.

This does not mean (nor did the authors intend it to mean) that the world will experience conditions lethal to all human life in this century, or even ever. They are saying that if the human race continues on its present trajectory, with modest and slow reductions in greenhouse gas emissions and no rapid deployment of other relevant measures like carbon dioxide removal and perhaps albedo modification techniques (reflecting sunlight), then within a decade or so we will be almost irrevocably committed to a pathway that will ultimately make the planet quite inhospitable to human civilisation. That is to say that we will be past the tipping points and *committed* within decades, even if we do not suffer extreme consequences for a much longer period. It is definitely an alarm call, but it is not a shriek of despair.

A Cascade of Feedbacks

What would a cascade of feedbacks look like? The earliest we are likely to see would start with the loss of the summer ice cover on the Arctic Ocean. This process is well underway, as the warming in the Arctic is three to four times faster than in the rest of the world. Since satellite records began in 1978, the *area* of ice left on the Arctic Ocean at the end of the summer melt season (early September) has fallen by an average of 40 per cent. And since the 1990s, the *volume* of the ice is down by more than half: when old multi-year ice (up to three metres thick) melts in a particularly warm summer season, it is now generally replaced by single-year ice, often less than one metre thick, that will melt quickly in subsequent summers. Shipping using the 'Northern Sea Route' across northern Russia and Scandinavia as a short-cut between the Atlantic and Pacific Oceans is already in the high hundreds of commercial vessels annually.

> As every year passes, we realise that the climate models, which do have some uncertainty, have all under-predicted the speed of the melting. Things are worsening at a much greater rate than anyone had thought, particularly in the Arctic. Three years ago, if you had asked me 'Should we hold the warming down artificially?', I would have said 'Look, I hope we don't have to do any of that geoengineering crap', but now I just can't see this predicament going in any other

direction. And I really hope we do proper, government-funded research on how these geoengineering techniques work, and how we can do them safely because, sure as eggs is eggs, we're going to have to do it.

Hugh Hunt, Centre for Climate Repair,
Cambridge University

A summer will come, perhaps within the next decade, when there is no ice left in early September except among the Canadian Arctic islands.[9] The Arctic Ocean will continue to refreeze in the winter for many more years, but the melt will come earlier each spring and the freeze-up later each autumn because the vast area of open water each summer will absorb and retain more heat. The loss of the sea ice and of snow cover on land (both of which bounce most incoming sunlight back into space) will speed the warming throughout the region – and that, plus the warmer water, will start the final melting of the Greenland ice cap. Warmer air will melt the glaciers from above, and as the altitude of the top of the ice cap drops, the air temperature at the surface of the glaciers will rise further: air temperature goes up one degree Celsius for every 350-metre drop in altitude. At the same time, the warm water lubricates the base of the glaciers, accelerating their slide into the sea. And unlike the Arctic Ocean ice, which is already floating on the ocean, the glacial melt-water raises the world's sea level.

The sea level would rise seven metres if all the Greenland ice cap melted, but that would take centuries. The more urgent question is whether the volume of fresh water flowing into the North Atlantic might once again destabilise the warming sea currents of the Atlantic Meridional Overturning Circulation (AMOC), better known as the Gulf Stream.

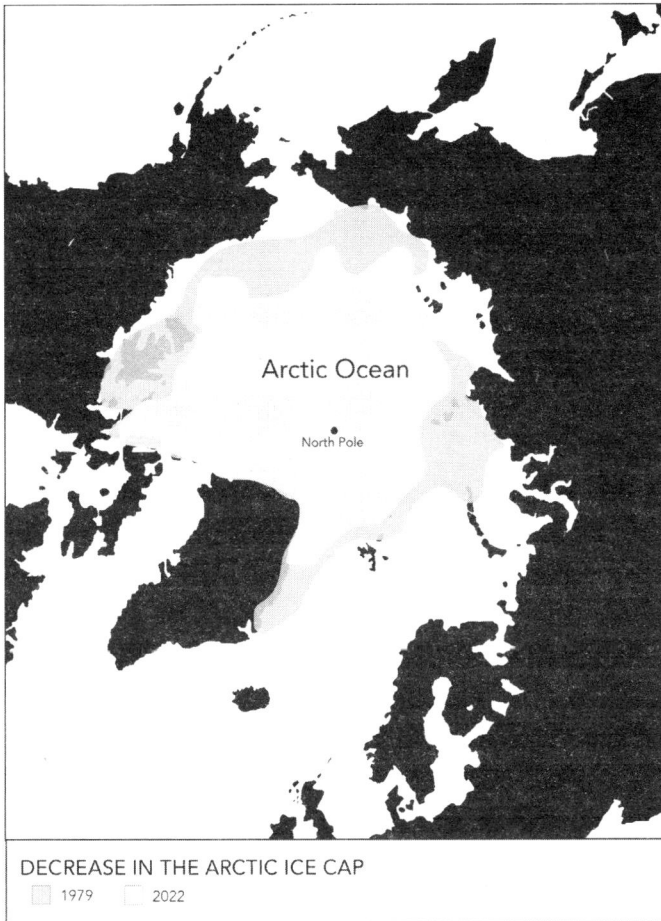

DECREASE IN THE ARCTIC ICE CAP
1979   2022

On one spectacular previous occasion, during the planet's emergence from the last major glaciation, the flood of glacial melt-water came from Lake Agassiz, centred on what today is the Canadian province of Manitoba, which covered an area as great as the Black Sea. The volume of water was large enough to raise global sea levels by at least one or two metres. It scoured out the valley of the Mackenzie River all the way north to the Arctic coast, surged east across the Arctic Ocean, and reached

the AMOC's 'overturning point' between Greenland, Iceland
and the United Kingdom. As a result the AMOC shut down,
causing a 1,400-year return to Ice Age conditions worldwide
between 12,900 and 11,550 years ago.

The moral of this story is that seemingly minor events (an
ice dam collapses on a big inland lake) can flip the entire global
climate into huge and long-lasting changes. It's not just sensitive; it's hyper-sensitive.

How Tipping Points Cascade

David Thornalley, an oceanographer at University College
London, studies the past behaviour of the AMOC in the hope
of understanding its future behaviour. I asked him what the
consequences would be now if the Overturning Circulation
stopped or slowed down drastically.

> They would be global in range. If there is an abrupt shut-
> down you would get widespread cooling over the Northern
> Hemisphere, focussed on the North Atlantic region: any-
> where from 8 to 10 degrees Celsius, and it could happen very
> rapidly. You can imagine just going into one winter and it
> being very, very severe. The sea ice could suddenly encroach
> down past Iceland, all the way down to off the UK, and that
> could happen over the course of a year or so. *That has hap-
> pened in the past.* [Author's italics]

Don't get the wrong idea. That wouldn't mean that global
warming had been cancelled. The AMOC would stop delivering
a huge amount of heat to north-western Europe, whose temperatures would swiftly drop to what you might expect in those
latitudes (Ireland is directly east of Labrador; most of Norway
is level with Greenland). But the planetary warming would

continue, and would probably overpower the local cooling after a generation or so. A roller-coaster ride in regional average temperatures is entirely possible.

The general view among climate scientists is that the AMOC (the Gulf Stream) is unlikely to stop completely in this century, although since 1950 it has already slowed down by 15 per cent and is likely to slow further (see Chapter 5 for an alternative view). In any case, the heating of the Arctic will continue, leading to further changes in the Jet Stream, intensifying boreal forest die-back, and, as we have seen, thawing permafrost. Any rearrangement of the Atlantic currents, even a minor one, might also have an undesirable impact on two other tipping points: the Amazonian rainforest and the West African monsoon. The knee-bone is indeed connected to the thigh-bone.

None of these events, if and when they occur, would be directly caused by human action. Once human greenhouse gases and other human activities have caused the initial warming, the rest of the changes are independent of human action – and unstoppable. That's how tipping points can cascade. The positive feedbacks that are the fine structure of these events self-amplify, and so great changes can come from relatively small inputs.

One last thing about tipping points: hysteresis. This is a term in physics which means that it's easier to go through a change in one direction than to come back the opposite way. If raising the temperature by two degrees melts the Greenland ice cap, it will take considerably more than two degrees of cooling to start the ice growing again. The cost of avoiding a flip in the state of some element of the climate system is almost always much less than the cost in time and energy of trying to bring it back once it has flipped – if that is possible at all.

* * *

We have ways forward which will work without terribly time-consuming or expensive further technological development. It's simply a matter of giving up our current teddy bears that we love to clutch, such as conventional hydrocarbons and fossil carbon fuels, and going into what we need to do to save ourselves.

Dennis Bushnell, chief scientist, NASA Langley Research Center

Dennis Bushnell was already feeling quite cross about the fecklessness of humanity when he said this in 2008. You may be feeling even crosser about it today. The outlines and scale of the emergency have been pretty clear for a quarter of a century now, and from the start the solution was as plain as day: stop burning fossil fuels. Not instantly and totally, because we live in a high-energy civilisation and in the 1990s didn't yet have adequate alternatives. But people were obliged to start developing and building those alternative sources as fast as they could, because they needed to be completely free of fossil fuels within fifty years maximum or there would be hell to pay.

And the damn fools didn't. They denied it, they faked it and they dawdled. Time passed and little was done, even though the science was getting better and the future was looking more frightening with every passing year. Now we are condemned to worrying about tipping points and feedbacks and all kinds of risky and expensive last-minute ways of stopping the warming before it utterly overwhelms us. How could they have been so stupid?

We'll discuss in detail in the next chapter what happened in recent decades (or rather, why it mostly didn't happen). In the meantime, consider this:

There's a place in Rochester where I can stand on a railroad track that looks down at the Erie Canal and over in the distance I can see cars on the interstate and overhead I can see planes landing at the airport. Four separate transportation and energy infrastructures put in place within the space of 150 years. They built the Erie Canal at an enormous cost in blood and money, and it generated a huge amount of wealth. Then the trains came in and they built a train line at an enormous cost and gave up on the canal. Then we built cars and we didn't use trains as much. Then airplanes and airports. Four different transportation infrastructures within 150 years. Switching infrastructures is not that big a deal. It's something we know how to do. We do it very quickly. In 1890, you could walk around and not see a single gas station. By 1920, there were gas stations everywhere. So within a few decades, there's no [climate] problem, if we're committed to switching infrastructures.

Adam Frank, Professor of Physics and Astronomy,
University of Rochester

He's right, of course. Switching the energy sources for an entire global society is a much bigger task than just changing one country's transportation network, but it is mainly a question of scale. Just get human emissions down fast enough that we never get close to 450 parts per million (ppm) of carbon dioxide in the atmosphere and the problem is mostly solved.

That's a daunting task, because we are already at 425ppm (mid-2023) and have been adding another 2.24ppm every year. But we don't have to accomplish it all at once. If we just make the easy cuts fast and shrink our emissions by half in the rest of this decade, then by 2030 we would only be putting 1.1ppm into the atmosphere each year. That would give us around fifteen years to eliminate the rest without breaching the 450ppm limit.

There is a wide and sometimes weird variety of ways available to achieve this, and many of them would even turn a profit in the end. From solar, wind, nuclear and geothermal power to electric vehicles, carbon-neutral octane fuels, biochar and passive houses, many solutions are ready. Others are farther away, but you can already make out their shapes: precision fermentation of bacteria to make 'farm-free' protein food, Direct Air Capture, Ocean Iron Fertilisation, and much more. Plus, of course, the existing low-tech standbys like planting trees, insulating houses and eating less beef.

It would all be entirely feasible if majorities of populations everywhere were willing to spend the money and accept the sacrifices that would be involved – money and sacrifices that would be quite modest compared to what populations expended in the last world war. I'm betting that you don't believe this is going to happen in the next seven or eight years: if you did, why would you be reading a book like this? Maybe later in the game, when the first big calamities are upon us, there will be a better response, and we must hope that won't already be too late. But cursing the human race for its folly won't help. Better to cultivate the perspective that Adam Frank has developed. It's useful, it's probably true, and ultimately you could even call it encouraging (although you shouldn't get carried away with that notion).

> Climate change is a generic consequence of building a civilisation.
>
> Adam Frank

Adam Frank is an astronomer by profession, but these days he prefers to call himself an exobiologist: someone who studies life on other planets, and in Frank's case particularly *intelligent* life and civilisations on other planets. Not that he knows of any or can even be sure that they exist. He studies *virtual* civilisations – but that can still be quite enlightening.

There is only one human civilisation, although it comes in a thousand different flavours. Frank thought it would help if we had a larger sample to work from, so he built mathematical models of 'exo-civilisations' that obey all the rules of evolution and population dynamics, and set them in biospheres that were more or less sensitive to the kinds of pressures that a high-energy civilisation would exert on them. His first conclusion, unsurprisingly, was this: 'If you start harvesting lots of energy from your planet and use it to do the work of civilisation-building, the planet is going to notice.' In other words, any planet that evolves life which then goes on to create a technological civilisation is liable to wind up in an 'Anthropocene' crisis.

Is Frank's experiment plausible? Yes, because the universe is so big that there could be a trillion civilisations in it right now and still not be a single one in our own Milky Way galaxy, but in a sense this doesn't matter. His models would still work, and still be useful, if there had never been another civilisation anywhere. Civilisations by definition are high-energy entities that alter their environments both deliberately and unintentionally, and in the great majority of cases they probably go on doing so until they find themselves in a quandary rather like our own. As their mastery over their environment grows, their population explodes. To feed that growing population, they take over more and more of the biosphere and divert it to their own purposes. Their energy usage expands almost exponentially as their technology advances, and that will have major planet-wide effects even if they don't have fossil fuels available. (If they do have fossil fuels available, it will just happen more quickly.)

Within 10,000 or 20,000 years of starting down the civilised road, these hypothetical intelligent species will be pushing so hard at the boundaries of the stable biosphere in which they evolved that they risk flipping the planetary environment into a new state far less favourable to them.

These extra-terrestrial alter egos are not gods. Like us, they will have evolved from pre-civilised ancestors who were not 'stewards of the environment', but wild animals living in a Darwinian dialectic with nature – which included other species of animals, and quite possibly other groups of their own species. Like us, they will have carried a good deal of that cultural baggage with them into their new civilisation, and they will probably end up breeding too fast and breaking too much. Undoubtedly, many will find that in order to survive they have to become 'planetary maintenance engineers' (as James Lovelock put it forty years ago) at least for a time. And some, of course, will not survive.

We had a bunch of parameters for the models that we built in. One of the parameters told us about the sensitivity of the planet: if you pushed on it [by using a lot of energy], how likely was it to roll off into a completely different state? We had a parameter for the civilisations that recognised what was going on and changed from a high-impact resource to a low-impact resource. We had a parameter for how fecund the civilisation was: how good was it at making babies? We had a bunch of these parameters and did lots of runs, changing the parameters to see how it would change the behaviours in the runs. And no matter what we did, the sequences would end up in one of three classes of behaviour.

The first class was what we might call the 'die-back'. The civilisation started harvesting the available energy resources and the population would rise very rapidly. The planet would begin to respond; the temperature of the planet would start to rise... The population would shoot past the carrying capacity of the planet, the temperature would warm to the point where that was causing more deaths, and then the population would drop pretty rapidly. Eventually it would level off, the temperature would also level off, and you'd reach an equilibrium, a sustainable state – at the cost of losing maybe 70 per cent of your population.

Of course, what's not in our models is the social response to losing 70 per cent of your population. If seven out of ten people died on Earth tomorrow, would we even be able to hold this civilisation together?

Adam Frank

This may well be the track humanity is on right now, because we're pretty close to the point where non-linear jumps in temperature occur. It isn't hard to guess which parts of our planet would bear the brunt of the die-back. But let's go back and consider Frank's other two kinds of outcome.

The next class of models was more dramatic. It was full collapse. The population would rise, the temperature would rise, and then you'd get yourself into some kind of runaway. The temperature kept rising and it got to the point where you couldn't sustain a population at all, and the population would just come down like a stone. Who knows what's left, but the population drops so rapidly that your high-tech civilisation is over. It's *Mad Max* – or you've gone back to hunter-gatherer. That's beyond our models. All we could see was that the population just came down really fast.

But there was one very disturbing phenomenon in the data: doing the right thing even a little too late may be of no use whatever.

This was depressing. When you ran models that switched from a high-impact resource to a low-impact resource, you could still have collapse even after you made that change. You could see the population rise, switch somewhere because they'd noticed it, and then it would come down and plateau for a while – but then drop like a stone. That was because of the non-linearity of the system.

The good news is that from the perspective of this physio-chemical-biological model, long-term civilisations on a planet are, in fact, possible. I was happy in the sense that we didn't get a zillion kinds of behaviour. We got three basic classes of behaviour.

Adam Frank's 'theoretical archaeology of exo-civilisations' may be no more than a grand metaphor, but his conclusions make us focus on the fact that our interactions with our own biosphere will decide our fate, and that we could lose everything if we don't get it right. We cannot know where we rank among the probable 'winners' and likely 'losers' on Frank's implicit tote board, but we should conduct ourselves as if both outcomes are possible. Because they are.

2

HOW COULD THEY HAVE BEEN SO STUPID?

A lot of people are putting a lot of effort into raising consciousness about climate change. Take Professor Hugh Hunt of the Climate Repair Centre at Cambridge University, for example.

HUGH HUNT: I like doing this with school kids. I usually bring along a little plastic bag which has something in it that looks a bit like poo. 'This is my poo from this morning,' I say. Sometimes I bring some scales along and I weigh it. 'Okay, that's 320 grams. Quite a big one. So let's say half a kilogram a day, 365 days a year, so we're talking about maybe 200kg a year of poo. That's my responsibility.'

I reckon my carbon footprint is about twenty tons of CO_2 a year, so my carbon dioxide emissions are a hundred times bigger than the amount of poo I generate. And we don't notice the CO_2, but you'd sure as hell notice if I produced a hundred times more poo than my normal daily amount.

GWYNNE DYER: So glad you don't.

HH: How do we get the message across that this CO_2 stuff isn't some kind of small little extra? It dominates our waste emissions by an enormous margin. What else do I do that even holds a candle up to the amount of CO_2 I generate? Nothing, nothing! And who tells us this? Nobody, because, well, we wouldn't want to jeopardise the economy, would we? That's why I do my poo bit – and there's always some kid who comes up to me afterwards and asks 'Is that really poo in that bag?' I tell them 'Oh, yeah. It is. Would you like to keep it?'

That's one way of doing it. Another is the traditional protest tactics, updated for a modern sensibility. Dr. Emily Cox's day job is at the UK Energy Research Centre.

A colleague of mine has a graph I love that shows how much people are concerned about climate change, and you get to April 2019 when Extinction Rebellion's pink boat was in Oxford Circus in central London, and all of a sudden the graph shoots up! There's this sudden jump in public concern on exactly that week.

When people say that protest tactics don't achieve anything, I show them this graph and say 'Look! From this we developed a public mandate, from that public mandate we developed a political mandate, from that political mandate we developed, among other things, a zero net emissions target which is now in government legislation, and now the government has to show how they are going to go about meeting it.' These movements might not save the world overnight and they have to work alongside a huge range of other partners, and everybody has to work together despite their differences, but we are seeing meaningful change as a result.

Emily Cox

The shift in opinion has been quite recent, but it is decisive. The biggest ever opinion poll on climate change, conducted by the UN Development Programme in early 2021, polled 1.2 million people in fifty countries and found that almost two-thirds (64 per cent) of them agreed that climate change was a 'global emergency'. The highest level of belief was in the United Kingdom and Italy (81 per cent). In Western Europe and North America as a whole it was 72 per cent, and every other region of the world came in at between 61–65 per cent. Even in countries that produce and consume very high levels of fossil fuels, most people strongly supported renewable energy (Australia: 76 per cent; the US: 65 per cent; Russia: 51 per cent).[10]

Major Disinformation

We can assume that the heatwaves, floods, forest fires and crop failures will continue to proliferate and intensify, providing regular confirmation that the climate emergency is real. The well-funded disinformation campaigns with which the fossil fuel industries try to deflect the public from backing policies that would damage their business models will continue, of course, but they are already switching from the old strategy of spreading doubt and disbelief to a subtler approach. Now they acknowledge that the heating is happening and that human activities are the primary cause, but they 'greenwash' themselves and try to shift the blame on to the public.

> Starting in 1908 with the discovery of oil in Persia, our story has always been about transitions – from coal to oil, from oil to gas, from onshore to deep water, and now onwards towards a new mix of energy sources as the world moves into a lower carbon future. [Note 'lower carbon', not 'zero' carbon.]
>
> BP website (June 2023)

The oil, coal and gas conglomerates don't care what the public believes as long as it doesn't support government action to regulate them, cut their subsidies or even tax them more heavily. (Shutting them down is not yet on the table.) So we have had heavily promoted slogans like 'Beyond Petroleum' – by the company once known as British Petroleum, but now just BP – and a big push to persuade the public that they have a new, 'clean' energy source called hydrogen. And it's true that it is clean by the time it is burned, but it's the variant called 'blue hydrogen'.

The person who coined the term 'blue hydrogen' has remained anonymous, but he or she was likely hired by BP, which was one of the oil and gas majors that established the Hydrogen Council in 2017. This is 'a global CEO-led initiative of 132 leading energy, transport, industry and investment companies with a united and long-term vision to develop the hydrogen economy'. The blue hydrogen they are promoting is really the natural gas they have been producing all along, some fraction of which they now propose to 'reform' by breaking the methane molecules into hydrogen that they can sell as a 'clean' fuel, and a far greater mass of carbon dioxide that they promise to bury underground (eventually, if they can find the space). It's not exactly a scam, but it is certainly intended to mislead people. And it says nothing about oil, which they have no plans to stop selling.

BP is only one of about a dozen 'oil majors', half of them state-run, that dominate the world market. It's the second largest of the old 'seven sisters', the non-state-owned oil companies, and its self-declared greenhouse gas emissions in 2020 were 374 million tonnes of CO_2, around 1 per cent of total anthropogenic (human-caused) emissions in that year. It has pride of place in the greenwashing hall of fame because of its brilliant propaganda success since 2004 with its 'personal carbon footprint calculator'.

The effect of this tool was to transfer the moral and political responsibility for those huge emissions from the producers to the consumers. The message seemed to be: 'It's not our oil wells but your dietary and travel choices and other personal behaviours that are the problem. Concentrate on fixing those, measure your progress obsessively, pat yourselves on the back for every tiny success, and leave us alone to get on with pumping the oil.' To many well-intentioned people it might have seemed as though the oil industry, or at least the BP part of it, had changed sides. It hasn't.

Like the old denialism, these subtler tactics are part of an industry-wide strategy of managed retreat. It involves a good deal of scorched earth, as such strategies always do, and in this case it leads ultimately to inevitable defeat: either those seeking to end fossil fuel use eventually shut the industry down, or out-of-control climate change shuts the entire society down. But each successive generation of senior executives since the industry first became aware of the climate problem in the 1960s has only needed to kick the can down the road for another twenty years to build their careers and make their piles of money. Why were we stupid enough to let them get away with it?

What might have been (if we were a different species)

When we discount the welfare of future generations in favour of those who are alive now, when we ignore dangers that aren't very close yet (and might never arrive), when we mistrust people from outside our own group and prefer comforting fairy tales to grim realities, we are displaying a mindset that was pro-survival for hunter-gatherers who lived in a dangerous and unpredictable world and could expect to be dead by their mid-thirties.

A longer lifespan and a lot more knowledge have improved us, no doubt, but we are still essentially the same people, and although our behaviour is not hard-wired, more often than not we follow the default emotional pathways that were laid down in a distant and very different past. It is plausible (though not necessarily true) that we might have ignored climate change, taken refuge in comforting falsehoods, and postponed action just as long even if the fossil fuel companies had not spent all that money on lying to us. They were merely protecting their own interests, they'll be safely dead before the bill comes in, and to hell with anyone who isn't sitting around their camp-fire. Typical humans.

Typical behaviour in any intelligent species that has only recently built a technological civilisation, Adam Frank might say. We cannot know which of his three categories of virtual civilisations we most closely resemble, and we should obviously act on the assumption that we still have a chance of being among the survivors, but it's too late now to take advantage of some of the softer options. By 1990, the scientific evidence for global warming was already strong enough that prudent governments would have started taking action on emissions, though nothing too dramatic yet. In fact, if they had acted then, we would never have had to do anything very dramatic.

Imagine a global but quite modest mitigation-only effort begun in 1990 – say, 2 per cent emission cuts per year for the developed countries and no limits on emissions in the first ten years for developing countries like China. That would have done for an opening round: it would have caused very little pain and it would have created incentives to work on new non-fossil energy technologies.'

Then in the second decade, 2001–10, the rich countries would continue with their 2 per cent annual cuts while the growing economies of the developing world (now including India) would accept not cuts, but a limit of 5 per cent on the annual growth of their emissions. Again, little or no pain, even for booming China, and if they complain loudly then settle for a 6–7 per cent annual limit. And the first large-scale wind farms and solar arrays would be springing up all over the place.

Finally, in the third round, 2011–20, the rich countries stick with their annual 2 per cent reduction in emissions while the cap on emissions growth in developing countries is tightened to only 3 per cent a year. Result: the rate of annual growth in global emissions has more than halved, and the amount of carbon dioxide in the atmosphere is 390ppm, not 420ppm. Humanity still has a lot of work to do, because there is still a net growth in CO_2 of about 1ppm a year, but we now have sixty years to winkle out the last net emissions, and we'll never have to go through the never-exceed 450ppm and experience a full 2°C of warming. Nobody took a big hit on their economy, and the wild weather does not start in the early 2020s. In fact, it probably never starts.

Now, I'm not saying that could really have happened here. I'm just saying that a somewhat wiser, rather less divided and turbulent global civilisation could probably have done it. We know that we are not that more advanced civilisation because we didn't do it, although we had all the evidence we needed. *And we didn't fail just because some oil companies lied to us.* We missed all the early, easy exits because of who we are.

That doesn't necessarily mean that our civilisation is doomed, let alone that the entire human race is, but it does make our road longer, harder and much more expensive. We are a species that has changed its behaviour greatly in the past – consider the move into the mass societies, for example – and given enough time we can do it again. Time is the issue, however, and it's not clear whether we can change at the pace that is now required. We don't even know exactly how fast that change has to be, but it's clearly a lot faster than we have been moving so far.

3

'STABILISED EARTH'

'Stabilised Earth' is not a comforting concept, but it's the best option on offer from the 'Trajectories' scientists. It sounds like having to walk a tightrope, balancing and re-balancing on every step, from here to eternity. Is it really that bad already? Well, here's a reality check. Are they saying that we have *already* left the Holocene for good? No more Ice Age ever? Yes they are – and that was a done deal 5,000 years ago.

A word of explanation. The Holocene is the name for the current interglacial period, around the thirtieth since the Ice Age began 2.6 million years ago. It began about 11,000 years ago with the usual big leap in average global temperature, taking the planet up even beyond the present warmth. In the typical interglacial, the temperature would then very gradually make its way back down over a period of 10,000 or more years, but the trend would always be downwards – and by now we would be reaching the threshold of the next major glaciation. Re-entry into one of these big, 100,000-year glaciations is always slower and gentler than the exit from the previous one, but according to palaeoclimatologist William Ruddiman, in the normal course of events:

> there would today be permanent snow cover, ice sheets growing over north-eastern North America and around the edges of the Arctic Ocean... It would be nothing like the full glaciation that existed 20,000 years ago, but we would be in the start of the next Ice Age.

This is the pattern Ruddiman expected to see when he began working with the data from the ice cores: a steady fall in the

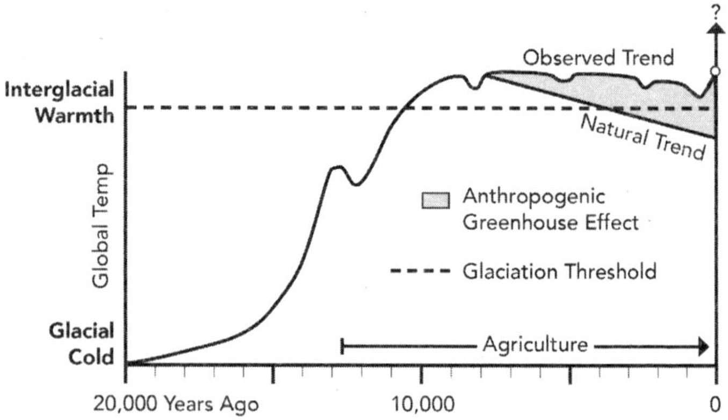
?
Observed Trend
Interglacial Warmth
Natural Trend
Global Temp
Anthropogenic Greenhouse Effect
Glaciation Threshold
Glacial Cold
Agriculture
20,000 Years Ago
10,000
0
William Ruddiman's Early Anthropocene Hypothesis

amount of methane in the air, indicating a steady fall in temperature. And that's what he found in the data for the early Holocene – but around 5,000 BP ('Before Present' – around 3,000 BC) the methane emissions started going up again. It didn't make sense, unless...

The First Farmers

Ruddiman boldly proposed that the cause of the about-turn in methane emissions might be human activities. Of course, he encountered strong opposition to this idea *(Who ever heard of human beings influencing the climate? The fellow must be mad!)* However, he started looking at carbon dioxide and found the same answer: the amount of CO_2 in the atmosphere, which soared at the start of the current interglacial, also declined in the expected manner until 5,000 BP – but then it too turned around and started rising again.

The 'Ruddiman Hypothesis' argues that by 5,000 years ago human farmers in the Middle East, Europe and Asia had cut down enough trees to send CO_2 levels soaring. The paddy fields of Asia were also contributing large amounts of methane to the mix. His conclusion: the global cooling trend was halted as a result of (inadvertent) human intervention. Steve Vavrus of the University of Wisconsin–Madison collaborated with Ruddiman in his later work on this phenomenon.

STEVE VAVRUS: There were so few people on Earth at that time. How could they possibly have had a global impact? But it's a bit like the tortoise and the hare race: the 'tortoise', the tiny ancient farming population, very gradually building up extra amounts of carbon dioxide in the atmosphere over thousands of years, may well have had a climate effect comparable to the 'hare' – the rapid increases in carbon because of industrialisation over the past 200 years. The time-scales are so vastly different that the rate of carbon input into the atmosphere doesn't have to be equivalent at all.

GD: How much warming did the early farmers cause?

SV: Based on the behaviour of the previous interglacial, the carbon dioxide level in 1850 should have been around 245 parts per million [without human intervention]. Instead, it was around 285ppm, so around 40ppm difference. Based on our climate model experiments, that would have caused an artificial warming of just under one degree Celsius globally, which is about the same as the observed global warming that industrial countries have caused over the past hundred years or so. Though the time-scales are very different, the magnitude of the climate change is similar.

GD: Would we be seeing 're-glaciation' now on a big scale if our ancestors hadn't started deforesting and farming the land thousands of years ago?

SV: Glaciation in the Northern Hemisphere is driven by the so-called orbital conditions that determine how much solar radiation we get, and the natural orbital conditions right now are favourable for an expansion of glaciers. We are at a minimum of radiation and have been for the past couple of millennia. I'm not suggesting that we would be seeing massive ice sheets all over the Northern Hemisphere were it not for early farming. Our climate models say that it would have been what we call 'incipient glaciation', so we would be beginning to see some of the ice sheets and glaciers expand as a response to this reduced summertime heating from solar energy. And we think that process was short-circuited by the artificial pumping of greenhouse gases into the atmosphere from early agriculture.

All unwitting, the human race became geoengineers 5,000 years ago: the next major glaciation was cancelled not by us today, but by our distant ancestors. They meant no harm by it; no more than beavers damming ponds. It's not even clear that they did any harm, unless cancelling major glaciations is a bad thing for the Earth System as a whole. What it does mean is that when we started burning fossil fuels in a big way around 1850, the average global temperature was already a full degree higher than it would have been without 5,000 years of agriculture. The message is that the Earth System is so tightly coupled that even a modest number of early farmers, given time, can push it into major changes.

Planetary Stewardship

We have already kicked the Earth out of the glacial cycle and now the big question is: can we stabilise the planet in a sort of super-Holocene where the average global temperature is only two or three degrees higher, and can we still somehow manage? Two degrees would have many disruptive consequences, but the world would still look at least similar to the

Holocene world. Or will we be pushed out of that by internal feedbacks?

Hans Joachim Schellnhuber

When other climate scientists read the 'Trajectories' paper, there was an almost audible sigh of relief, because finally somebody was saying it aloud. And here comes the 'planetary stewardship' bit again:

> In essence, the Stabilised Earth pathway could be conceptualised as a regime of the Earth System in which humanity plays an active planetary stewardship role in maintaining a state intermediate between the glacial-interglacial limit cycle of the Late Quaternary and a Hothouse Earth. We emphasise that Stabilised Earth is not an intrinsic state of the Earth System but, rather, one in which humanity commits to a pathway of ongoing management of its relationship with the rest of the Earth System.
>
> From 'Trajectories of the Earth System in the
> Anthropocene'

That sounds like a tricky balancing act. It also sounds like the unenviable job description for James Lovelock's overwhelmed 'Planetary Maintenance Engineer' as rewritten by a panel of HR experts. But the meaning is clear. Whether governments recognise it or not, whether the public is ready to hear it or not, the emergency has arrived, and it will still be filling the horizon well beyond the span of our own lives.

> We will already have reached the 'Hothouse Earth' state at three degrees Celsius warming, because after that we will just continue sliding, we just cannot halt – the equilibrium

might be a thousand years away, but it is a drift that cannot be halted.

A 'Stabilised Earth' is a recognition that we have changed so much on the planet already that there is no way back to the virgin Holocene state, the state of the planet that kicked off modern civilisation as we know it. Instead, we have to accept that either we drift off towards a 'Hothouse Earth', or we try to find a manageable state, a state where we stay within planetary boundaries, do not cross tipping points and they do not add up to a planetary threshold.

It is a stabilised Earth where through our active management we collectively become planetary stewards and manage the system. We recognise that we are now in the driving seat, we are the dominant force on Planet Earth. The global human enterprise exceeds any volcanic eruption, any earthquake, even solar radiative changes. All these natural variations continue happening, but we are now the biggest force.

This is not a naive thought of some kind of return to a Garden of Eden, but recognising that we have no choice, we are simply so big and so dominant that we now need to drive the vehicle. Currently we are just sitting there and not really recognising that we are the ones with the levers now. We are starting to understand how these levers work, but we are not using them, and it's time to use them.

Johan Rockström

I don't think we're in Kansas anymore, Toto.

If we aren't in Kansas, where are we?

One of the odd things about the climate change world is that 'everybody knows' the next bad news years before it is confirmed. So, for example, various officials at the COP28 climate conference in Dubai in late 2023 were still talking about keeping the average global temperature under 1.5°C at least until the mid-2030s, although 'everybody knew' this target was no longer possible. They even managed to ignore the scientific paper published just before COP28 by Robin Lamboll and Joeri Rogelj of Imperial College London which explicitly said that the 'Remaining Carbon Budget' for 1.5°C would run out in 2029.[11]

The Remaining Carbon Budget (RCB) is how much more carbon dioxide you can emit before the global average temperature reaches a certain level, and at the beginning of 2023 the RCB for 1.5°C was 250 gigatonnes. Since human beings are emitting forty gigatonnes each year, it doesn't take a mathematical genius to work out that the budget runs out in 2029, so why was everybody pretending that it would last until 2035?

> **ROBIN LAMBOLL**: Unfortunately, that was probably generous even at the time. I think that we've been hinting that this was not really accurate before then, but yes, the IPCC moved slowly. And you know, it has a sort of great weight to it. It takes a long time to consider all the evidence.
>
> **GD**: What are the practical consequences of arriving at 1.5°C in 2029 rather than arriving six years later?
>
> **RL**: To be honest, if we're not actually reducing emissions very much anyway, it doesn't really matter. Current government action is not compatible with this goal anyway.

Cynicism is a defence against despair. You can't blame Dr Lamboll at all.

* * *

So where will we all be in the mid-2030s, which is barely far enough away to qualify as 'the future'? We may already be in a world where massive super-hurricanes are pushing ten-metre 'storm tides' up river estuaries, especially shallow, funnel-shaped ones facing the Western Pacific, the Bay of Bengal, and the US east coast. Baltimore and Washington will have taken big hits already, and Bangkok, which had never before been touched by a typhoon in its short history, will be struck by a mega-typhoon that drowns the entire central city. The Gulf states, already the hottest inhabited part of the world, will be losing population as temperatures climb even higher and those who can afford to leave move out. The Mediterranean countries, the US Southwest and much of Africa will not be far behind. And almost all the world's coral reefs will be dead or dying.

In this scenario, the global average temperature in 2035 would be +1.6°C, but some of the bigger non-linear feedbacks that we thought would only be arriving after we reach 450ppm would be starting to stir. The writing would be on the wall: we'd be effectively committed to a minimum of 2.5°C of warming, and it's a safe bet that this would commit us irrevocably to the 'Hothouse Earth' pathway.

So what do we do?

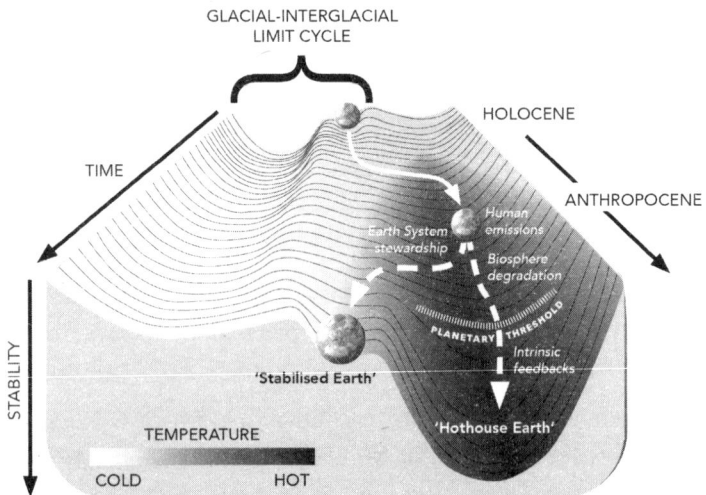

This is the other diagram that Marten Scheffer brought to the key 'Trajectories' meeting in Stockholm, showing a 'stability landscape' in which the pathways along which the Earth's climate might evolve are depicted as valleys. Our past emissions mean that Planet Earth is already too far down the valley leading to Hothouse Earth to return to the real Holocene, but it still has a chance to carve out a new pathway back up the valley-side that would deliver us to a 'Stabilised Earth' that is pretty close to Holocene conditions. 'It's more of a cartoon than a mathematically correct diagram,' as Will Steffen admitted, but it demonstrates why it will be an uphill task to reach a place of relative safety.

To keep the planet from plunging ever deeper into the valley labelled 'Hothouse Earth' (with Hell at the bottom) and steer towards 'Stabilised Earth' – still hotter than now but less catastrophic than our current destination – we must maintain the planet in Holocene-like conditions, with a temperature rise no greater than 2°C above pre-industrial (a 'super-Holocene' state).

Creating that last-chance 'Stabilised Earth' would require 'deep cuts in greenhouse gas emissions, protection and enhancement of biosphere carbon sinks, efforts to remove CO_2 from the atmosphere, *possible solar radiation management* [my italics], and adaptation to unavoidable impacts of the warming already occurring...,' as the 'Trajectories' authors put it. In other words, multi-trillion-dollar global investments in huge engineering projects of several kinds, and enormous changes to industry, business, consumption, employment, lifestyle, taxation – pretty well everything really, all done in a tearing hurry, in order to stabilise our planet in a condition worse than the most extreme climate conditions we have yet experienced, but hopefully still bearable in most places. And with no guarantee that doing all this will succeed in creating their 'Stabilised Earth', because we have left it very late.

THINGS TO DO IN A CATASTROPHE

Greenhouse Gas Emissions

The 'Trajectories' analysis is not necessarily a death sentence, but it certainly sounds like a life sentence of hard labour. Is it realistic to expect that people can really act with such discipline over the long term?

Cutting greenhouse gas emissions was once touted as the only thing we had to do to solve our climate problem (although that was never true), and in the public discourse it retains pride of place. But in the past thirty years, the proportion of energy that the world draws from coal, gas and oil has remained almost unchanged at around 80 per cent of total energy use. Renewables (11 per cent) and nuclear power (4 per cent) have grown apace, but overall demand for energy has grown so fast, mainly due to the rapid development of some non-Western countries, that carbon-neutral energy sources have made virtually no inroads into the fossil fuel share of global energy use. There may be a significant acceleration in carbon-neutral energy getting underway now, and, in the longer term, fossil fuels can and must be eliminated, but it is unlikely that we are going to see cuts of half or more in greenhouse gas emissions over the next decade. Not impossible, but politically implausible.

Carbon Sinks

Next on the 'Trajectories' list: 'protection and enhancement of biosphere carbon sinks'. This is definitely the flavour of the moment, and has been since the Paris Accord was signed in 2015. Because most of the people there didn't really believe that emissions cuts were going to come fast enough to stave off disaster without a lot of help, they signed up for an unspecified amount of what is known as Carbon Dioxide Removal (CDR). This can be as simple as planting trees or restoring swamps and peat bogs so they can soak up more carbon dioxide from the atmosphere, but it also includes much more technological alternatives, such as Direct Air Capture (extracting CO_2 directly from the air with large machinery and sequestering it in underground rock strata and/or in the deep oceans).

In principle there is nothing objectionable about most of these proposals, but the trees, swamps and bogs all involve land use issues, which means money, lawyers and delay, while the high-tech ones are heavy engineering projects that would be expensive, take quite a while to get going, and even longer to produce significant results. In the end, no doubt, Carbon Dioxide Removal will be needed on a very large scale, but it's unlikely to have a large impact on the state of play by the early 2030s, which is probably our deadline for deciding how to avoid taking the 'Hothouse Earth' pathway. More on this later.

Solar Radiation Management

'Solar Radiation Management' (SRM) is meant to reduce the amount of incoming sunlight by a small amount in order to hold down the heating at the planet's surface. Among other methods, this is to be accomplished by spraying aerosols: for instance, reflective sulphur dioxide particles into the stratosphere, or seawater droplets into the air to 'brighten' marine clouds just above the ocean.*

SRM incites some people to fury and worries even most of its advocates, because it involves a major intervention in an extremely complex planetary system – 'Gaia', no less – of which we have only a limited understanding. On the other hand, SRM

---

* There have also been proposals to put giant mirrors into space and similar heroic engineering feats, which nobody studying SRM takes seriously but the opponents of SRM love to dwell on.

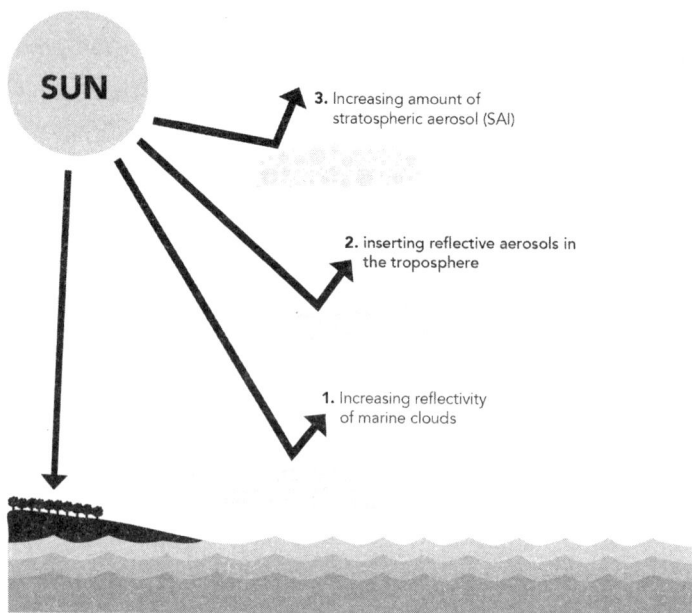

is relatively cheap and fast-acting compared to almost all the other options, and might just be the temporary bridge we need to transport us from the unsustainable present to a viable future without a mass die-back in between. There is much to be said about it – both pro and con.

Adapting to the Unavoidable

Well, yes, of course: everybody will have to do that. But if we're not also doing much or all of the above already, then we will probably find adaptation to our new environment quite difficult, if not impossible.

If we look at the 'Nationally Determined Contributions' volunteered by all the nations that signed the Paris Climate Agreements, the best-case scenario is more than three degrees of warming. And let's face it, those are only voluntary, nonbinding 'commitments'; they don't have the kind of legal teeth enforcing them that the World Trade Organisation does on trade exchanges. So we'd be optimistic to think we were going to limit the warming to three-point-something degrees – and that's without factoring in some of the harmful tipping points or feedbacks that might make matters worse.

A pragmatist would have to say four degrees is looking distinctly possible later in this century, but that's a deeply dangerous place to go. As a species we live in a very constrained climate niche and have done for thousands of years, and even if we just take the three degrees of warming as a reasonable, almost optimistic best guess, over three billion people fall outside the climate niche we live in now and they'll fall outside it on the hot side. They'd be experiencing mean annual temperatures of above 29°C, that hardly anybody on the planet anywhere experiences now. A natural, logical thing for many of those three billion or so people to do would be to move away from inhospitable climate conditions, and sadly we don't seem to have the humanity in us to embrace that scale of migration.

Tim Lenton

We can't really take the human story onwards from here in any coherent way. There would still be billions of people alive when the planet hits +3°C (on its way to +4°C and on up, for all the feedbacks will have been triggered by then), but the human race would already be in mass die-back territory. Whether what's left at the end is a much-diminished civilisation of chastened survivors or just 'a few breeding pairs' is unguessable, and in any case this is not a cautionary tale. We can, however, say something about what the planet would look like beyond that point, with or without human survivors, for no calamity that human civilisation could impose upon the biosphere could outdo the ones that have already happened without our help. In that respect, the mass extinctions of the past are quite instructive.

4

HOW BAD COULD IT GET?

It started off as a joke. Back in the early 2000s people were looking at abrupt climate change in the past, and there was this very sharp blip in the record around 55 million years ago called the Paleocene-Eocene Thermal Maximum (PETM). Nobody knew what it was, but you could see that there was massive perturbation to the carbon cycle, massive global warming, some extinctions, a big increase in erosion. At a workshop, I put a list on the wall of what happened during the PETM and what we think is going to happen to the climate now – and it was basically the same list. And as a joke I said: 'Well, what if the cause is the same both times?'

Gavin Schmidt, Director, NASA Goddard Institute for Space Studies

There was no asteroid strike, no 'igneous province' erupting for many millennia to explain the PETM, but there was a very fast rise in the CO_2 level and a 5–8°C jump in average global temperature. Could some previous species on Earth have evolved intelligence, built an industrial civilisation, and then killed itself off by wrecking the climate with its emissions? After all, most things that can happen once on a planet can also happen more than once.

It was just idle speculation in 2004, but a decade later the Kepler space telescope had already discovered 2,000 planets orbiting other stars. The search for life on those planets was underway, and Adam Frank visited Gavin Schmidt in New York to discuss what kind of technological 'signatures' in the light from those planets might indicate that one was home to a high-energy industrial civilisation. Schmidt mentioned the puzzle of what had happened on Earth in the PETM, and their study expanded to include what kind of 'technosignatures' a non-human civilisation that had arisen on this planet 55 million years ago, industrialised, and ultimately died from runaway warming driven by its own industrial emissions, would have left behind.

They immediately discovered what specialists on ancient climates already knew: as you go back in time, the evidence gets sparse very fast. Ice cores will take you back 800,000 years at most. Ocean sediments will take you back as far as 200 million years, which is plenty for the PETM (in fact, that's how we know about it), but they are quite limited in the kinds of information they preserve. And don't imagine that you'll find a physical artifact that has survived this long, such as a fossilised Silurian* wearing a fossilised wristwatch.

In the end, Schmidt and Frank wrote a proper scientific paper about it, concluding that there was no evidence that an earlier intelligent species had caused the PETM.[12] 'I would say it was unlikely,' said Schmidt, 'but perhaps not as unlikely as people who just dismissed it out of hand before they thought about it would have said.' And having had their fun, they returned to their real work.

I think it's very unlikely too, so why am I bringing it up here? Well, mainly because it shows that we are not as great a threat to the biosphere as we imagined. Going extinct would be a disappointing outcome for the human race, but in the PETM we have an episode where, with or without Silurians, the planet

---

* 'Silurian' was the name Gavin Schmidt gave to the hypothetical first civilised species on Earth, who hypothetically wiped themselves out in the PETM. He took the name from a *Doctor Who* episode that had terrified him as a kid. It featured an ancient race that was awoken by nuclear experiments in a coal-mine.

experienced a degree of warming and associated disruption comparable to what we can expect if we let our own climate crisis go all the way to +6°C, and there were only scattered extinctions except in the deep oceans. Full return to 'normal' appears to have happened within 200,000 years, and the event is even associated with the appearance of a number of new mammalian species – including the first primates.

I'm not even insisting that the hypothetical Silurians would have been made extinct by the warming. Maybe they would have just lost their civilisation and dwindled back to a small population of hunter-gatherers (or the Silurian equivalent) who continued for hundreds of thousands of years until some other change killed them off. The average length of time a large mammalian species like ourselves lasts in the current Cenozoic Era is between 1 and 2 million years.[13] We only have 300,000 years on the clock, so maybe a human population one-thousandth of the present size could enjoy a long post-civilisation afterlife as hunter-gatherers.

Mass Extinctions

The real 'mass extinctions' were much bigger events than the PETM. There have been only five of them in the past half-billion years. They wiped out between 75–95 per cent of all living species, and one thing that they all have in common is that human beings couldn't have caused them. Nor, even if they were to occur now, could we prevent or control them.

> Things are always going extinct in the background, which is why 99.99 per cent of anything that was ever with us on the planet is gone, but ever since the dawn of animal life on our planet about 540 million years ago, diversity has been increasing... Coral reefs take off, trilobites, the typical

invertebrates – and then we have a big freezing event about 444 million years ago that ends up killing things even in the tropics. This is the first of five big mass extinctions that we can see in the fossil record.

Jessica Whiteside, School of Ocean and Earth Science, University of Southampton

The Ordovician mass extinction was probably caused by the colonisation of the land by plants. As they spread, the plants would have drawn down immense amounts of carbon dioxide from the atmosphere, cooling the planet so much that there appear to have been glaciers quite close to the equator. But that was a one-off event, irrelevant to our current concerns.[14]

The most recent mass extinction, 66 million years ago, was the 'K-T Event' that ended the Cretaceous period and ushered in the Tertiary period. The 'event' was the impact on Earth of a very large asteroid (ten kilometres in diameter) that wiped out half the planet's living things in a single hammer blow and took down most of the rest in an 'impact winter'. Like a 'nuclear winter', the huge amounts of dust and smoke boosted into the upper atmosphere by the asteroid impact blocked most sunlight for many months or a few years. The plants died and the animals starved to death, but there's no point in our worrying about asteroid strikes that we cannot prevent. Maybe one day, if we acquire the ability to divert such a threat…

The real source of concern is the three intervening mass extinctions, all of which occurred at times when there was a high concentration of carbon dioxide in the atmosphere, apparently due to very large and long-lasting volcanic eruptions. Human civilisation is currently putting carbon dioxide into the atmosphere ten times faster than those mass-extinction-triggering volcanoes. Could we be the volcanoes this time?

Earth Formation
4.5 billion years ago

Cambrian Explosion
541 million years ago

Ordovician Extinction
450 million years ago

Late Devonian Extinction
375 million years ago

Permian-Triassic Extinction
252 million years ago

Triassic-Jurassic Extinction
201 million years ago

Cretaceous-Tertiary Extinction
66 million years ago

Current

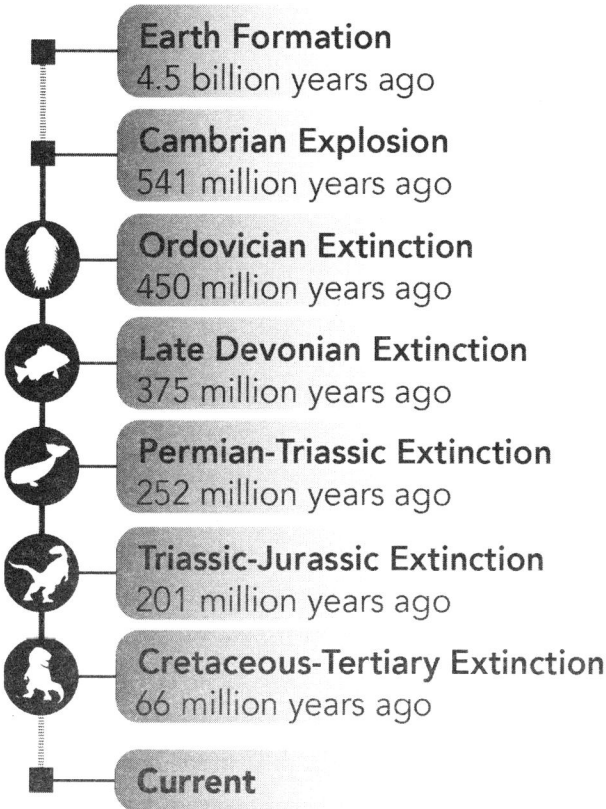

Mass extinctions

These three mass extinctions – the Devonian (375 million years ago), the Permian (252 million years ago) and the end-Triassic (about 200 million years ago) – are quite similar. They were driven by volcanic activity in a series of pulses that continued for up to a million years, with lengthy quieter periods between the pulses. The immense amounts of lava mostly came in the form of 'basalt flooding' that covered large areas of land or sea floor, and enough carbon dioxide was released to drive the global average global temperature up by as much as 14–16°C. A 'kill mechanism' effective enough to cause a mass extinction, however, requires something more than just very hot, very wet conditions.

Recent research suggests that the kill mechanism in all three extinctions was hydrogen sulfide. Ocean temperatures were very much higher, and warmer water can contain less dissolved gas – in this case, less oxygen. Moreover, as the temperature difference between the poles and the equator shrank drastically, both the winds and the ocean currents slowed down, because what drives them is precisely that difference in temperature.

The currents may never have shut down entirely, but in a world with not much by way of temperature differences, the volume of water they were moving dropped dramatically. The 'overturning' circulation in which huge masses of cold, dense, highly oxygenated surface water sink to the bottom like a thousand Niagaras slowed greatly or stopped. The oceans stratified, and, without the constant replenishment of oxygen from above, the deeper ocean became anoxic – no oxygen.

Anoxic oceans allowed anaerobic sulphur bacteria to emerge from the seabed muds where they had been hiding most of the time since the oceans became fully oxygenated, probably between 600 million and 800 million years ago. Green and purple sulphur bacteria multiplied, producing hydrogen sulfide (H_2S) as a waste product, and that gradually killed off almost all the oxygen-breathing life in the oceans. Finally it reached the surface to devastate life on land as well, by two methods: clouds of H_2S that poisoned animals directly and higher clouds of the same gas that destroyed the ozone layer and exposed all land-based life to potentially lethal doses of ultraviolet radiation. And this was not a one-off event. It could last for decades or centuries, and then go away for millennia – only for the whole catastrophe to be replayed (with an already diminished cast of living players) as the massive volcanic eruptions restart and the temperature begins climbing again.[15] (The above is the hypothesis of Lee Kump of Pennsylvania State University, and it matches the facts pretty well.)

> **LEE KUMP**: It's been more of a challenge to demonstrate what the effect was on land than it has been for the oceans, but it's clear that some predominant organisms, both plants and animals, went extinct at this time. There's some evidence that there was mass devastation to forest ecosystems. [The end-Permian] was the major extinction of insects, which is interesting if we're thinking about poisonous atmospheres, so there's no doubt that there was a mass extinction on land.
>
> **GD**: How does the current warming for which we are largely responsible compare with the mass extinctions we've just been talking about?
>
> **LK**: There are different ways to think about this. There's the absolute magnitude of the warming and CO_2 rise, and then there's the rate of change. The rate of warming at the end of the Permian was maybe ten times slower than the modern rate, but the absolute amount of carbon dioxide released was probably significantly larger than we could release even if we were to combust all of our fossil fuels. The warming of the Permian is likely to have been much larger than anything we would likely do over the next few centuries through fossil fuel burning. It could have been 15°c or perhaps even more, and persistent for longer periods of time.

That's probably the answer. The volumes of carbon dioxide being released into the atmosphere now and then are very different. The amount of CO_2 that had to be released into the atmosphere to raise the average global temperature by 16°C in the End-Permian (100,000 billion tonnes) was *forty* times greater than the amount of all carbon accessible in modern fossil fuel reserves, plus all the carbon already burned since the industrial revolution.[16]

It was 'new' carbon, released from far underground, that drove the End-Permian extinction, whereas we are just burning or otherwise releasing CO_2 that is already stored somewhere on the planet's surface. There is a finite supply of that: probably enough to drive the average global temperature up by 6°C if we burn all the fossil fuels, but not a lot more. We might be able to do another PETM and make noticeable inroads into the catalogue of living species, but we are not volcanoes. We are not even in the same league. Mercifully.

5

IT'S MORE SENSITIVE THAN WE THOUGHT

When I began the research for this book three years ago, I shared the widespread assumption that countries would go on failing to meet their targets for emission cuts, and that this would lead to a warming crisis in the mid-2030s. That would be the point at which desperate people would want to use artificial interventions in the atmosphere to hold the heating down and win more time for emissions cuts. Therefore, I argued, we should be actively debating and researching geoengineering options soon, in order to make well-informed decisions then.[17]

The revelation in 2023 that we were going to run through the last of our Remaining Carbon Budget in 2029, not 2035, certainly lent urgency to that conclusion, but it didn't really change it. Most climate scientists had already reached the private conclusion that we were destined to exceed the 1.5°C limit by 2035. They just didn't want to discourage the troops by a premature declaration of defeat on that front.

Then, in November 2023, a climate scientist called James Hansen proposed a radical re-reading of the evidence – one that implies a very different schedule. Hansen argued that the debate should start in earnest *right now*, and actual geoengineering within a few years if it proves safe and feasible.

> The 1.5 degree limit is deader than a doornail... and the scenarios you would need to stay under two degrees are just imaginary.[18]
>
> James Hansen, Director, Program on Climate Science,
> Awareness and Solutions of the Earth Institute at
> Columbia University

It wasn't the first time Hansen had had the dubious honour of delivering bad news about climate change to the world. In 1988 he had the job of explaining to the US Congress and the American public that global warming was a real threat. He and his fellow scientists did such a good job that there was serious action on the climate front for a few years (the foundation of the Intergovernmental Panel on Climate Change, the signing of the UN Framework Convention on Climate Change, etc.) until the fossil fuel industries mobilised to thwart further progress.

Now Hansen is back, all these years later, to warn us that the Equilibrium Climate Sensitivity (ECS) is 50 per cent higher than the number that the IPCC and most climate scientists have been working with. The ECS denotes how much warming you will eventually get when you double the amount of carbon dioxide in the atmosphere – and if Hansen is right, it's a very big deal.

The ECS has always been a hard number to pin down, but the global consensus for some time now has put it around 3°C (with ±1.5°C after it denoting much uncertainty). But it gave people a common baseline, and a good deal of the mathematical modelling of possible climates incorporates this number. But Hansen's explosive paper claims that the ECS is really 4.8°C, again with the inevitable ±1.5°C after it. If that's true, then the best-case outcome is higher than +3°C, and the worst case is higher than +6°C.

That would put us in the same climatic zone as the Pale-ocene–Eocene Thermal Maximum.

Hansen also claimed that there is another 5°C of warming 'in the pipeline' in the form of 'slow feedbacks' that will straggle in after decades, centuries and even thousands of years. A disheartening prospect, but he does point out that our descend-ants are not doomed to suffer all of that extra 5°C of heating if they invest in enough Carbon Dioxide Removal techniques and technology to remove the excess CO₂ from the system before it produces that warming. (They will still curse us, though.)

* * *

What gave Hansen the confidence to make these bold claims was that another scientist had come up with a new and better way of determining the Equilibrium Climate Sensitivity. Scientists trying to model ancient climates had been getting answers that were all over the place, mainly because they couldn't pin down how the clouds worked. (That's the hardest part of modelling modern climate too.) It might be better, the logic went, to use actual physical evidence rather than virtual models.

The atmosphere is 'well-mixed', so the carbon dioxide level is the same everywhere. It's therefore easy to figure out from ice cores what the carbon dioxide level was at the 'Last Glacial Maximum' (LGM) – the coldest centuries of the last major glaciation, around 20,000 years ago. Since there are always ice cores available in the polar regions even for warm periods, it's just as easy to work out how much *more* CO₂ was in the atmosphere 7,000 years ago.*

This gives us a good long run of increasing carbon dioxide, with enough time for all the slow feedbacks to come in: 13,000 years in all. If we can match that with the change in average global temperatures over the same period, then we can figure out the Equilibrium Climate Sensitivity at last. Unfortunately, you need to know the average temperatures in a number of different regions over the land and the sea to work out the 'average global' temperature.

That's not so hard for 7,000 years ago. There are still ice

---

* 7,000 years ago we were already in the warm 'Holocene' interglacial, but human beings had not yet started putting any carbon dioxide of our own into the atmosphere to muddy the picture. Carbon dioxide levels were higher 7,000 years ago than at the LGM, mainly because warmer oceans can contain less dissolved CO₂ and return some of it to the air.

cores for the frozen parts of the planet, and there's pollen, diatoms, coral, all kinds of 'proxies' for the temperature in the unfrozen parts. For 20,000 years ago we had ice cores for the (much larger) frozen parts of the land surface, and various forms of sea life as proxies for the ocean temperatures, but nothing useful for the temperature on the land surface in the lower latitudes, nearer to the equator, where there was no permanent ice. That was the missing piece of the puzzle – and then a graduate student at the University of California San Diego called Alan Seltzer came up with the solution. He called it a 'noble gas paleo-thermometer'.

Noble gases (neon, argon, krypton, etc.) do not interact with the environment chemically, but they do dissolve into groundwater, and different isotopes are absorbed at different rates depending on the temperature at that time. So Seltzer collected samples of ancient groundwater from local monitoring wells in the San Diego area (he was at the Scripps Institute at UCSD), took them back to the lab, and analysed them. After making allowance for ice sheet growth, he determined that the land temperature in southern California 20,000 years ago averaged a full 7°C colder than it is now. For the world as a whole, it was 6°C colder.[19]

Whoa! Nobody saw that coming! This changes everything!

Well, no, not everything, but it certainly does change our figure for Equilibrium Climate Sensitivity. Climate scientists were working with an estimate of 3°C. Now they are learning that the same amount of carbon dioxide released into the air (that hasn't changed) will actually give us a lot more warming than we thought. For Hansen, it was confirmation of what he already suspected and feared, and he ran with it.

After making allowances for this, that and the other thing in the usual scientific way, Hansen concluded that the ECS is probably 4.8°C, not 3°C – and a great deal flows from that. First of all, there's much more warming in the pipeline than we thought. Five degrees is what you get with the 'fast' feedbacks; by the time the 'slow' feedbacks also straggle in, it's ten degrees. That's mass death, your civilisation gone, and sea levels 25 metres higher. Now, this would all take a long time to happen – centuries – and we could slow and eventually halt the warming process by stopping our emissions and doing a lot of CDR, but ten degrees higher is now permanently on our agenda as a possible long-term outcome.

Secondly, although the biggest damage done by a much higher climate sensitivity lies in the future, it has serious consequences for the present as well. Hansen says that the speed at which the climate is warming has increased by half in the past decade, from 0.18°C per decade before 2010 to 0.27°C now.

> We're already at +1.2° degrees, and I argue that within the next year or two (2024–25) we will go up to 1.6°, 1.7°C warming. Then after the El Niño finishes, it will drop back – but it's not going to drop all the way back to 1.2°. It may go to 1.4° or so, but because the planet is so far out of equilibrium it's not going to drop much. So we're already at sort of the 1.5° warming, and we've got two degrees C in the pipeline by some time in the middle to late 2030s is my estimate.
>
> But we really can't do that, so I think during this decade, during the next several years, we're going to have to get people to understand that *we have to minimise this geoengineering of the planet, which is what we have been doing.* [Author's italics]
>
> James Hansen

There's a rhetorical trick in play here. Hansen is about to recommend that we do various forms of SRM in order to avoid disaster, and he knows that many people flinch if they even hear geoengineering mentioned. So he's trying to take the edge off the proposal by pointing out that we have already been geoengineering the planet for a very long time.

We had already unwittingly cancelled the next major glaciation 5,000 years ago, long before we even knew realised that we were living in an Ice Age (Ruddiman's Hypothesis). Then we had the industrial revolution about two hundred years ago and started doing a lot of the really bad kind of geoengineering (dumping greenhouse gases into the atmosphere), and we wouldn't have a warming problem now if we hadn't done that.

However, at the same time we also started putting a lot of particulate matter into the air (dust, dirt, soot, smoke and tiny droplets of liquid) that blocked or reflected incoming sunlight and cooled the planet. These aerosols did huge damage to people's lungs and to the natural world, but for the first 150 years of the industrial revolution, say until about 1950, the invisible greenhouse gases that caused the warming and the highly visible pollution that reflected incoming sunlight and caused the cooling more or less balanced out, so there was no noticeable change in the average global temperature. After that the balance began to tilt, as industrialisation spread into the developing countries (warming) and industry in the developed world began to clean up the pollution (more warming).

By now we have a net warming nearing 1.5°C with much more to come. Yet it is still rare to encounter a climate scientist who is comfortable with discussing the need to deliberately manipulate the two kinds of geoengineering, warming and cooling, that we are engaged in every day. Both the warming geoengineering (greenhouse gas emissions and the like) and the cooling geoengineering (brown clouds, ship tracks and the like) are being done by us – but up to now nobody in the field has been doing double-entry bookkeeping on a regular basis. In fact, they were scarcely keeping track of the cooling geoengineering at all, although they knew it was there.

I first encountered the bizarre yet intense reluctance of climate scientists to address geoengineering issues directly fifteen years ago, when I was interviewing Hans Schellnhuber, the founder and then the director of the Potsdam Institute for Climate Impact Research. He is an excellent scientist and an honest, friendly, normally quite straightforward man. Nevertheless, when I began venting about the lack of visible panic in the scientific community as we all cruised serenely towards disaster – journalists often do this sort of thing, in the hope of shaking something loose in the interviewee's carefully composed facade – he replied in a curious, almost coy way.

> We might have a little bit of help from our dirty friends, namely the aerosols in the atmosphere. You know that they have this global dimming, masking effect. It's almost ironic: without these aerosols we would probably have already much higher global warming. So it may turn out that if we would do a very subtle management of aerosols by sulphur filtering in China, India and so on, in line with carbon dioxide reduction, we might still save the day; but it may become a very tricky game, actually.
>
> Hans-Joachim Schellnhuber

Schellnhuber was a good man in a bad time. Most scientists and many of the journalists who talked to them knew that government and IPCC policies on climate were deplorably inadequate, but the first priority of the most senior scientists was to preserve the existence of the institutions they led, while hoping for better times. That's why so many interviews at that time resembled amateur dramatic productions of 'The Emperor's New Clothes', with both the scientists and the interviewers aware that the former could not afford to bring the whole delusional facade crashing down and so were obliged to speak almost in riddles.

At that time Schellnhuber was the climate change adviser to Germany's Chancellor Angela Merkel, the only leader of a G7 country who was a fully qualified scientist. She may well have shared his conclusions, but she would have been equally constrained by the realities of contemporary German politics. What he said to me then was probably a good indication of what the leaders of the developed world, or at least the better-informed ones, were already thinking in private in 2008.

They didn't believe any more that mere mitigation – just cutting GHG emissions – would happen fast enough to keep us below the 'never more than +2°C' threshold, but the idea was that the new coal-fired power plants and smokestack industries of Asia would pour enough aerosol pollution into the atmosphere, especially sulphate particles, to hold the temperature down even as Western polluters cleaned up their act. Then a decade or so later, when the Asian emitters started to clean their act up too, 'a very subtle management of aerosols' might keep their sulphate emissions high enough to hold the average global temperature under +2°C until mitigation finally got the CO_2 concentration back down below the danger level sometime later in the century. Which would be, he admitted, 'a very tricky game'.

This never seemed a likely outcome, unless Asian parents care less about their children's lungs than Western parents do, but we will never get to test the theory because a different component in our makeshift aerosol shield seems to have come unstuck first. This is the 'ship-tracks', the emissions from the funnels of the 60,000 giant commercial ships that carry 90 per cent of the world's trade across the oceans – mostly container vessels and tankers that until recently burned 'bunker oil', the heaviest (and most polluting) distillate of crude oil.

In 2010 the International Maritime Organisation, with the best of intentions, began changing the rules about how much sulphur dioxide ships were allowed to emit. A decade later, in 2020, they radically cut the permitted amount of sulphur in marine fuel oil from 3.5 per cent to only 0.5 per cent. The 'ship tracks' that used to shade the major shipping lanes began to disappear, and a lot more sunlight began to reach the sea surface.[20]

Meanwhile, in 2013, China declared a National Air Quality Action Plan that has been extraordinarily successful. Pollution in general has been reduced by 42 per cent and sulphur dioxide in particular, spewed mainly from coal-fired power plants, by 87 per cent. This has spared hundreds of thousands of lives already, and is reckoned in the long run to lengthen the average Chinese life-span by two years. It has also raised China's average temperatures by 0.7°C in just ten years.[21]

China has reduced its aerosols in the past fifteen years, and aerosols from ships decreased especially in 2020. There's a great inadvertent aerosol experiment now ongoing... decreasing cloud cover and cloud brightness and thus increasing the sunlight absorbed by Earth, especially in the North Pacific and North Atlantic regions, where shipping is the source of a large fraction of the sulphate aerosols. The solar radiation absorbed by Earth [in those regions] has increased by about three watts per square metre, or a global average of 1.4 watts per square metre.

This increase of absorbed solar radiation is the reason that Earth's energy imbalance has almost doubled since 2015. A one-watt increase of absorbed solar radiation is equivalent to more than a twenty-year increase of greenhouse gases at their

current high rate of increase. That's why global warming will
accelerate.

James Hansen[22]

No doubt this seems like enough bad news for one day, but
Hansen has one more grim prediction to offer: that the North
Atlantic and Southern Ocean Overturning Circulations will
shut down due to a flux of fresh melt-water from Greenland
and Antarctica in this century, and possibly as early as mid-century. And if the AMOC shuts down, it's goodbye to the relatively
temperate climate of Western Europe.

So what does Hansen suggest that we do? One: carbon taxes
everywhere to speed the transition out of fossil fuels. Two: Solar
Radiation Management measures at any altitude from sea level
to the stratosphere to hold the heat down while the transition
continues. Three: financial and technical aid to developing countries so that they can afford to choose non-fossil energy options
to grow their energy supply, even though they have abundant
local fossil fuel resources that are cheaper to exploit and would
provide more employment. A short list, but a good one.

* * *

I feel that this latest contribution from Jim and his co-authors
is at best unconvincing. I don't think they have made the
case for their main claims, i.e. that warming is accelerating,
that the planetary heat imbalance is increasing, that aerosols
are playing some out-sized role, or that climate models are
getting all of this wrong. And I certainly don't think that
they've made the case for engaging in potentially disastrous
planetary-scale geoengineering projects.

Michael E. Mann[23]

The response of the climate change 'community' to Hansen's declaration of a climate emergency was led by Professor Michael Mann of the University of Philadelphia, who functions as the unofficial gate-keeper of the current climate orthodoxy. This reaction will have a lot of resonance elsewhere, as Hansen has a reputation for being a maverick who defies the collective wisdom of his peers: 'very much out of the mainstream,' as Mann puts it. On the other hand, Hansen also has a track record for being often and surprisingly right, and a dozen other reputable scientists collaborated on his paper. His predictions are more extreme than those of any other leading climate scientist: that current policies will probably take us permanently past +2°C average global temperature by the mid-to-late 2030s – and up to +5°C within the next century.

If Hansen is right, we need to start taking emergency measures almost immediately, so we can't afford to wait for all those predictions to be confirmed or disproved by rival scientific papers in the usual time-consuming way. That's why Professor Hugh Hunt, Deputy Director of the Centre for Climate Repair at Cambridge University, and Dr Robert Chris of the Open University made a special plea for a rapid resolution of the controversy.

We find Mann's rebuttal of Hansen's claims unconvincing. Crucially, Mann does not engage directly with Hansen's analysis of new data [regarding paleo-temperatures and various other matters].

Hansen projects that in coming months, lower levels of aerosol pollution from shipping will cause warming of as much as 0.5°C more than IPCC models have predicted. This will take global warming close to 2°C as early as next year (2024), although it is likely then to fall slightly as the present El Niño wanes.[24]

This prophecy provides us with a useful early check on the accuracy of Hansen's predictions – at least those that deal with the lost cooling effect due to reductions in human marine and land sulphate emissions. A consensus on his predictions about the ECS, the annual rate of global average temperature increase, and the persistence and scale of slow feedbacks even after greenhouse emissions reach zero, will take longer to reach. But it must not take too long, or we may find ourselves trapped irretrievably on the one-way track to 'Hothouse Earth'.

PART TWO

ALL ABOUT EMISSIONS

... in which it is discovered that despite a cornucopia of potential remedies for the current emissions crisis, few can be made available at scale in time to get us off the hook

6

ENERGY AND EMISSIONS

I plotted the data from some predictions of future carbon dioxide emissions that were made in 1990, so that's thirty years up to 2020. That model made assumptions about GDP, population growth and the amount of carbon dioxide in the atmosphere. It has not been wrong by one part per million of CO_2 [in any year] for the past thirty years, which I find incredible.

Thirty years ago, they were able to make these predictions that have proved to be remarkably accurate because they know how the human psyche works. We've all been very much driven by economic growth, and GDP has a really strong correlation to energy demand and energy use. It's been really depressing.

James Haywood, Professor of Atmospheric Science,
University of Exeter

So far, not so good, but we have to keep going. This chapter and the next are about how we might avoid a very ugly outcome by reducing and ultimately ending our emissions of greenhouse gases. This one concentrates on the 76 per cent of emissions that come from generating energy.

One might think that changing from fossil fuels to alternative energy sources would be a relatively straightforward process, since most customers don't care where their electricity comes from so long as it's there when they flip the switch. On the other hand, they do care how much it costs, and changing most of the energy infrastructure of a high-energy global civilisation tends to be expensive.

There are only two ways to cut emissions when dealing with energy: either replace the fossil fuel source of the energy with a different, non-emitting source; or reduce demand so that the energy is not needed. Cutting demand obviously includes taking steps to reduce one's own personal emissions, but lifestyle changes matter less than political action or society-wide economic and technological choices. During the pandemic year 2020, for example, when hundreds of millions of people were forced to work from home, thereby drastically reducing their personal carbon footprints, global CO_2 emissions fell by only 7 per cent – and by the end of that year monthly emissions had returned to pre-pandemic normal.

Experience teaches us that efforts to reduce demand, however desirable this outcome might be, do not yield large or rapid results. Indeed, most experts assume that demand for energy will continue to grow more or less in step with the global economy – which is forecast to double in size by 2050. In this case we are working to a closer deadline of 2035, but it would be foolish to expect less than one-third growth in demand by then.

> We're not getting rid of fossil fuels. We're getting rid of the subsidies for fossil fuels, but we're not getting rid of fossil fuels for a long time.
>
> Joe Biden, *The Hill,* October 2020

If we want a realistic answer to the question of how much countries can cut their greenhouse gas emissions by 2035, we need to look at the sources of non-fossil energy. These include the classic 'renewables' – solar, wind and water power – but also nuclear and geothermal power. Most of this new non-fossil energy will be produced as electricity, but some will be used directly to produce heat, especially in steel-making and cement production, as we shall see. And still in the exploratory phase or the earliest stages of development are fusion power and the various attempts to make hydrogen a useful fuel ('blue' and 'green' hydrogen and ammonia).

The key question we must put to each of these techniques is: how much of the current fossil load can we expect them to displace by 2035? Pessimism would seem advisable, since despite the rapid growth of renewables, until recently they had not displaced *any* fossil fuel production. The rapid expansion of the renewable sector in the last decade has been just fast enough to compensate for the failure of nuclear and hydro power to grow significantly during that time, but fossil fuels have retained their 81–85 per cent share of primary energy consumption.

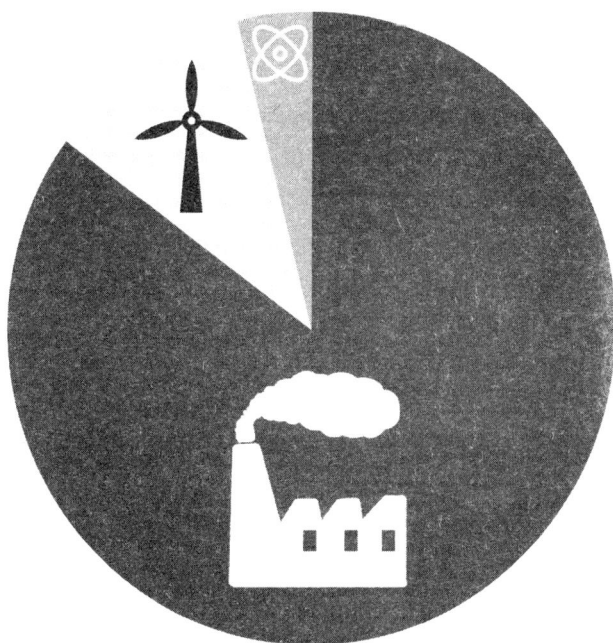

● Fossil fuels 85%   Renewables 11% ● Nuclear power 4%

So here follows a look at all the non-fossil fuel options that are operational or at least under development, with estimates of how much fossil energy they could displace by 2035.

THE BIG THREE
Photovoltaic Power (aka Solar Panels)

Photovoltaic (PV) cells are unquestionably the star performers of the past decade. In many markets, the cost per kilowatt-hour of power from solar panels has plummeted by 85 per cent, so 'grid parity' (i.e. with the cost of fossil power) has now been achieved. China is the largest producer of solar panels, and also has the largest installed PV capacity. Solar panels accounted for about 14.7 per cent of global power generation in 2023.[25]

In the past decade, the doubling time of photovoltaic power has been between three and four years. It is very unlikely that that growth rate can be maintained, but it is plausible that PV power might, by 2035, account for around 35 per cent of *current* global consumption, displacing a comparable proportion of coal and gas production. It should be noted, however, that in the intervening years the global economy will probably have grown by 30–35 per cent, and that there are around 600 new coal- and gas-fired power stations under construction.[26]

Potential obstacles to more rapid deployment include short-ages of so-called 'minor metals' like cadmium, gallium and tellurium that are used in newer thin-film panels. There is also a rapidly growing requirement for energy storage, since a day-time-only power source that supplies up to one-third of global electricity consumption requires an alternative source of power when PV is unavailable. Nevertheless, this is clearly one of the two great hopes for retiring enough fossil fuel power to have a chance of staying under +1.5°C in the mid-2030s.

Concentrated Solar Power

Once the leading solar technology, Concentrated Solar Power (CSP) uses arrays of mirrors to heat molten salt and drive turbines to make electricity. It has fallen behind its spectacularly successful photovoltaic (PV) rival because the initial capital cost is higher, but in regions with strong year-round sunshine it is competitive over the longer term. Since it stores its heat, it can continue to provide electricity for hours after sunset. The installations tend to be very large, but they do good service in the desert areas of Morocco, Spain and the United States.

Wind Power

With a decade's head start on solar, beginning in the early 2000s, wind power now accounts for 5 per cent of global electricity production. China, once again, is the biggest producer and has the largest installed wind capacity. Windmills may be preferred over solar panels in countries in the higher latitudes, where the sunshine is weaker and less frequent, and wind does have the advantage that it also works in the dark. However, a new phrase, 'wind famine', has entered the language, because the wind sometimes ceases to blow for quite long periods. It dropped so steeply over the North Sea in September 2021 that the United Kingdom was able to draw only 7 per cent of its electricity from wind instead of the usual 25 per cent.[27] Partly to compensate, very large wind farms are being built at a considerable distance offshore, where winds are stronger and more reliable.

* * *

Solar power is growing much faster than wind, so it will eventually be carrying more of the load – and the combined growth for these two leading renewables, if it continues at the current pace, could lead to a 50 per cent share for renewables in total global electricity consumption by 2035, even allowing for a one-third expansion in the global economy.

That could put the no-more-than-1.5°C goal just within reach, provided that no tipping points are crossed, and some other specific conditions – such as at least a 30 per cent reduction in methane emissions – are met. However, renewables will not be able to go that high until and unless we have solved the storage problem.

Storage

Fossil fuels will produce energy instantly whenever and wherever it is required, but renewables are only available when the sun is shining or the wind is blowing. Since energy input and energy output must always balance on the grid, there will be times when energy demand is so high that grid managers are obliged to burn some fossil fuel to match it. At other times, when demand is low and the grid cannot use all the renewable energy that is coming in, they have to throw the surplus away. What is needed is a means of storing surplus renewable energy until the high demand is there, hours or even weeks later.

In the past, short-term storage to meet sudden surges in demand was done mostly by traditional 'pumped hydro storage', in which two lakes or reservoirs, one near the top of a hill and another at the bottom, are connected by a tunnel with a turbine at the bottom. Water is pumped up to the higher pond when demand for electricity is low, then released in less than a second when demand peaks to rush back down the tunnel, drive the turbine, and generate more electricity to balance the grid. Pumped hydro storage provides instant response and very large capacity, but it requires hilly terrain and a good deal of earth-moving. In the United States, for example, there are 228 gigawatts of wind and solar power, but only 22 gigawatts of pumped storage.

New short-term (4–12 hours) storage to meet soaring demand is provided mostly by batteries similar to the lithium-ion batteries in electric vehicles, coupled together in large enough numbers to meet grid requirements. Research is underway to develop comparable batteries that are lighter or use less rare or cheaper materials: sodium-ion batteries, solid-state electrolyte batteries, lithium-sulphur batteries, and even a version of inventor Thomas Edison's original nickel-iron battery of 1901 (which can produce hydrogen as a by-product). Some non-battery ideas for short-term storage are also being explored, such as Edinburgh-based Gravitricity's proposal to hoist 12,000 tonnes of iron weights up 1,000-metre deep mine-shafts with excess electricity, and then recover the energy as the weights come back down (25 megawatts for three hours: enough to power around 8,000 homes) or Toronto-based Hydrostor's $1 billion project in California to dig a million-cubic-metre space deep underground to store highly compressed air in off-peak times and then release it through turbines when demand peaks (400 MW for eight hours).

Long-term energy storage (days, weeks or even months) requires a different approach, and the rival technologies for doing that are still in the pilot-plant stage or even earlier. They will be needed to cope with long-term fluctuations in the renewable energy supply due to seasonal change, wind famines, monsoons, etc. – the kind of problems we were once able to solve by just burning more fossil fuel. Unfortunately, getting large amounts of energy back out of long-term storage in the form of electricity is still an evolving technology. It can be done, but the 'efficiency' – how much electricity you can get out as a proportion of how much you originally put into storage – is running at 45–65 per cent. This is unimpressive compared to 70–85 per cent for pumped hydro storage and 80–90 per cent for lithium-ion batteries, and it may explain why China, which plans to add ninety-seven gigawatts of new energy storage by 2027, depends almost entirely on the latter technologies.[28]

Nevertheless, all the Western industry leaders in long-term energy storage (LTES) are using versions of 'pumped thermal electricity storage' technology. They employ different mediums to store the energy – the Danish firm Stiesdal uses crushed basalt rock heated to 600°C; the US-based Malta Inc. (spun off from Google X in 2018) uses molten salt; and Britain's High-view Power, the only one with a grid–scale renewable energy storage facility already under construction (near Manchester), just uses very hot water – but all three are working with heat pumps to maximise the output of electricity. A relatively fast roll-out of these systems should be possible, because almost all the technology involved is straight off the shelf.

Adequate storage is critical for a rapid transition to a world economy with a high reliance on renewable energy. If it becomes a bottleneck, the chance of getting half our electricity from renewables by 2035 falls to zero.

Space-Based Solar Power

The notion of putting solar panels into orbit and beaming power back to Earth was always attractive in principle, because it obviated the storage problem. Solar 'farms' in geostationary orbits would be in full sunlight almost all the time, and would beam the power back to Earth at radio frequencies unaffected by cloud cover or other terrestrial weather phenomena. However, the cost of boosting large amounts of mass into such orbits (35,786km high) was prohibitive – at least until the advent of reusable rockets (such as SpaceX's 'Falcon Heavy' and 'Starship', or Blue Origin's forthcoming equivalents) transformed the economics of orbital flight.

The European Space Agency calculates that the break-even

point for putting solar farms in space is $1,000 per kg. The current cost of using Falcon Heavy would be $1,500 per kg, but Starship, once fully operational, might be able to do it for $200 a kilo. That still leaves a great many questions to be answered, from the mechanics of robotically assembling 2,000 tonnes of lightweight solar panels into an array 1.7km in diameter, to housekeeping details like how to replenish the reaction mass necessary to keep the whole thing in exactly the right orbit. (Electrical propulsion from the satellite's own solar array would reduce the amount of reaction mass needed, but it would eventually would run out.) Nevertheless, half a dozen countries are now taking this technology seriously.

In 2023 Northrop Grumman Corp. and Caltech carried out a first experiment in successfully transmitting high-frequency radio waves from their 'Space Solar Power Demonstrator' mini-satellite to a ground station that turned them into DC electricity. The British government commissioned a report from the Frazer-Nash Consultancy that envisions an operational 2-gigawatt orbital solar power station by 2040. China plans to generate one megawatt of energy from space-based solar panels by 2030 (with a commercially viable solar space station by 2050). However, none of this offers any credible prospect of displacing a significant amount of fossil fuel by 2035, so we can pass on to matters more relevant to our deadline.[29]

Nuclear Power

Nuclear power can provide very large amounts of electrical power without any greenhouse gas emissions, and it's ideal for providing base-load power in a post-fossil-fuel system that depends mainly on wind and sunshine for its power. The technology is mature and reliable: only one accident has caused a major loss of life (forty-three deaths at Chernobyl) in operating many hundreds of reactors over a period of almost seventy years. Nowadays in most places it is seen as a potentially useful tool in the effort to reduce emissions – but not, curiously, in large parts of Europe.

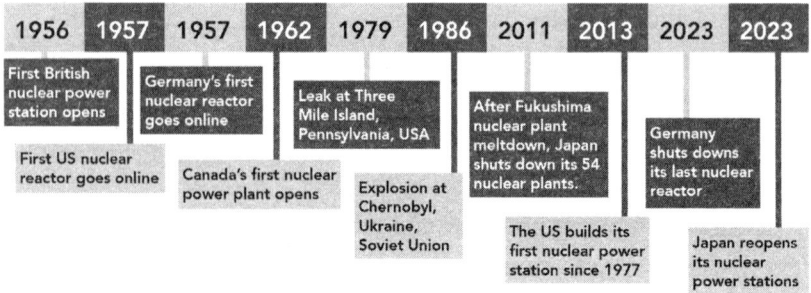

1956	1957	1957	1962	1979	1986	2011	2013	2023	2023

1956 First British nuclear power station opens

1957 First US nuclear reactor goes online

1957 Germany's first nuclear reactor goes online

1962 Canada's first nuclear power plant opens

1979 Leak at Three Mile Island, Pennsylvania, USA

1986 Explosion at Chernobyl, Ukraine, Soviet Union

2011 After Fukushima nuclear plant meltdown, Japan shuts down its 54 nuclear plants.

2013 The US builds its first nuclear power station since 1977

2023 Germany shuts downs its last nuclear reactor

2023 Japan reopens its nuclear power stations

Why do so many people abhor nuclear power? Carbon dioxide emissions are the greatest threat to human civilisation, and nuclear power had huge potential to reduce them. Today there are 438 commercial nuclear reactors in the world, supplying about 10 per cent of the world's electricity. If the Green movement had not turned against nuclear energy, there would have been three or four times as much nuclear power by now, and the world would be nowhere near +1.5°C.

This was not a foregone conclusion.

The story begins in California in 1971, when environmentalist David Brower founded the Friends of the Earth, an organisation that was explicitly anti-nuclear power from the start. Much has been made of the fact that FoE's founding donor, Robert Anderson, was the owner of a major oil company (now trading as ARCO), and it was certainly in the interests of the oil industry to strangle its emerging nuclear rival in its cradle.[30] This does not necessarily mean that David Brower 'sold out', nor was Robert Anderson necessarily acting on behalf of the oil industry; he was a very rich man who was known for being a maverick and a bit capricious in his donations. Rather than indulging in conspiracy theories, it's probably best to think of the emerging Green movement and Big Oil as what Marxists used to call 'objective allies': no conscious collaboration was necessary, but the job got done. When Greenpeace emerged in 1973 and immediately joined the anti-nuclear power campaign, the startled and bewildered US nuclear power industry went on the defensive. It has been there ever since.

Chernobyl

Greenpeace's founding demonstration was against a nuclear weapons test, but it didn't bend over backwards to make the distinction between nuclear weapons and nuclear power. Both organisations were and still are happy to let their supporters confuse the two, presumably on the rationalisation that it is deceit in a good cause – and most people who know the difference forgive them (including me) because the other causes they support are, on the whole, good ones. But they certainly went into overdrive when the world's only nuclear accident that killed people by lethal radiation doses, the Chernobyl tragedy, happened in the former Soviet Union (present-day Ukraine) in 1986. Twenty-eight plant workers were killed right away, another fifteen people subsequently died of thyroid cancer, and an area of 2,600 square kilometres around the plant was declared an 'exclusion zone' to be entered only by guided tours. But if you knew the old Soviet Union in its last few years, you wouldn't even have found the incident surprising.

The Chernobyl reactors were built to a new design, but it was a seriously flawed one (there was never a proper exchange of safety information with other nuclear operators outside the Soviet bloc). The operators were not properly trained and the management was slovenly, which was par for the course in those days. The heroic plant workers who died preventing an even worse disaster were betrayed by a dying regime that even the reforming new leader Mikhail Gorbachev could not save. When he said that Chernobyl was 'perhaps the real cause of the collapse of the Soviet Union', he meant that it had made it impossible to ignore the conclusion that most people had already reached in private: that the system was corrupt, incompetent and moribund. I spent a lot of time in the Soviet Union in 1987–91 and you could hear open contempt for the regime in every kitchen conversation in the country; 'Chernobyl' had become almost a verbal shorthand for it.

Chernobyl was not a message to everybody from the gods to shut their nuclear industries down. Normal procedure when a dam bursts or a plane crashes is to investigate the technical flaws or human errors and fix them, not to close the entire industry down – and indeed the United States did not close its nuclear industry down entirely despite the Greens' best efforts. America still draws about 20 per cent of its electricity from nuclear plants built forty years ago or more, but investment in nuclear power was already slowing in the late 1970s because of the Green campaigns – and Chernobyl stopped it dead: no new nuclear power plant was begun in the US between 1977 and 2013. Germany and several other European countries did shut their reactors down – which is a large part of the reason why France and the United Kingdom, which kept theirs open, now have per capita emissions of greenhouse gases that are about half those of Germany.

Fukushima

The only place outside Europe and North America with a severe nuclear power allergy is Japan. There was always an understandable nervousness about all matters nuclear in Japan – the only country ever to be struck by nuclear weapons – but the country was drawing 30 per cent of its electricity from nuclear power before the Fukushima events of 2011. Nobody died or was even injured at the Fukushima Daiichi plant in 2011 (although 19,000 died as a result of the sub-sea quake and the tsunami that devastated the city). However, the Japanese media conflated the zero casualties at the Daiichi plant with the deaths from the tsunami, leading to the immediate shut-down of all Japan's 54 reactors.

They stayed shut for a decade, while Tokyo embarked on building twenty-two new coal-fired plants to replace them. If the surge in Japan's greenhouse gas emissions failed to move either the public or the government, the surge in energy prices after the Russian invasion of Ukraine did the trick. (Japan imports 99 per cent of its coal, oil and natural gas). By 2023 Tokyo had reopened seventeen reactors, and the government plans to build next-generation nuclear power plants in the 2030s.

The number of countries where fossil fuels have replaced nuclear power is small, and the extra amount of CO_2 consequently being emitted by these countries will not decide the fate of the planet. Elsewhere nuclear power is enjoying a modest comeback. There are fifty-four new nuclear power stations under construction, half in China and India, and even one or two in France, Britain and the United States. Nuclear power will continue to be part of the portfolio of energy sources we need to build a carbon neutral energy system. The question is: how big a part?

The answer will depend partly on cost, but also on whether the risk of running nuclear reactors, however much it has shrunk as the designs have improved, is low enough to justify their continued use far into the future.

Small Modular Nuclear Reactors

We take a salt and melt it. We dissolve the nuclear fuel into it and we pump it into the reactor and create energy. That energy turns into whatever you want – hydrogen or electricity – and it's very cheap. All that is not too magical. The magic in the salt is that when something goes wrong – not if but when, because some day, somewhere, it will – this salt, if it gets out of the reactor, solidifies and captures the radioactive elements so that they don't go anywhere.

As opposed to conventional nuclear, where the worst-case accident is a gas release, we would have a fluid release that solidifies into a rock. Technically it's a molten salt, but in your world, to all intents and purposes, it's a rock. It doesn't dissolve in water, it doesn't release any gases, and it only melts at 700°C. The 'worst case' accident scenario is still contained within the fence, which is dramatically different from the nuclear we have today. That leads to a lot of savings in complexity and cost, which means we can build them fast. And they're compact, so we can get to new regions.

Troels Schönfeldt, CEO, Seaborg Technologies, Copenhagen

As Troels Schönfeldt's reactors are also on barges, they can get anywhere in the world by sea – and whenever a new customer shows up, they play the Hallelujah Chorus.

The mood around nuclear has definitely changed, and Seaborg Technologies is not the only outfit promising cheap, safe power drawn from Small Modular Reactors (SMRs). They tend to come in modules of up to about 200 megawatts, each independent but designed to be linked up in larger numbers (up to twelve in one design) to provide whatever amount of power is required at the site. Being small(ish), they are intrinsically easier to cool and therefore safer. They can also be made in a factory rather than having to be constructed on site. This helps with both cost (actual production lines, almost) and with the amount of time required for construction – months or a year or two, not a decade.

It will be up to the regulators to decide whether these new designs are safe enough – and the regulatory process is so strict that not one of these new-model reactors exists yet, even in a pilot version. Then, once they're approved, it will be up to the market to decide whether it wants them.

> Our strategy is to get the cost down enough that it will happen by itself, because if something is cheap enough, it happens... If you look at the de-carbonisation targets that people are trying to meet [which are not ambitious enough], the scale is insane, so we're building this for scalability. We're designing the first reactor with the intention of going to a mass-produced unit. We're building on barges that we can sail to market, in existing shipyards that are high quality and efficient... We have partnerships with some of the biggest shipyards in the world and our target is to get to 200 barges a year by 2035. To put this into perspective, 200 barges would be approximately five to ten times the total offshore wind built to date in the world, so in a couple of years we would overtake renewables as a whole.
>
> Troels Schönfeldt

Why barges? Because SeaBorg reckons that the places where its nuclear technology will be most in demand, at least in the early years of the operation, will be in what used to be called the Doldrums and is now called the Intertropical Convergence Zone (ITCZ), between 500km north and 500km south of the equator. Winds are notoriously low or entirely absent and a thick band of cloud producing heavy precipitation and frequent thunderstorms encircles the equator, so neither wind nor solar power produces the benefits as elsewhere. Nuclear power would sell well in that environment.

> I sometimes say that we will be a hundred times the wind industry, and I think we could get there. We need to get there. But we need a hundred or even 1,000 companies to get to that level.
>
> Troels Schönfeldt

That may happen and it may not, but the man at least grasps the scale of the task.

OPPORTUNITIES AND DIFFICULTIES

Water Power

Hydroelectric power, mainly in the form of large dams, still accounts for one-third of total renewable power and 16 per cent of all electricity generation, but there is limited scope for expansion since most of the economically and technically attractive dam sites have already been exploited in Europe and the Americas. A recent study for the US Department of Energy, for example, identified 61 gigawatts (GW) of hydroelectric power potential on unexploited US rivers and streams, but anticipated that only 2 GW would actually be developed between now and 2040.[31]

> A deeper reservoir with less surface area is generally better for the climate and produces lower greenhouse gases than a reservoir that is wider and shallow. There's just a lot more vegetation at the bottom of a very wide reservoir, and when it decomposes in low oxygen waters it produces a lot of CO_2

and methane. Look for mountainous areas where you have these deep but narrow reservoirs. You want to maximise your electricity and minimise your emissions.

Ilissa Ocko, senior climate scientist,
Environmental Defence Fund

Hydroelectric power is not unequivocally 'clean' power. It depends on the emissions from the reservoir, which tend to be higher in tropical regions. China, India and Africa still have considerable potential for hydro development, but the environmental cost in terms of flooded land is high and in some cases so is the emissions cost. Nevertheless, of the so-called 'clean' sources of electricity, only wind and nuclear have lower emissions per kilowatt-hour.

Tidal power – damming the entrance to ocean bays that have a large tidal range, and exploiting the power of the water as it surges into the bays and back out again in a daily or twice-daily cycle – is in its infancy, with only two plants in operation – in France and South Korea – and a few more in development. Exploiting open ocean currents or waves is even further back in the queue, with prototype machines only now going into the water. All the ocean options also risk being very expensive because salt water is extremely corrosive. None of them offers the prospect of a major jump in 'green' energy production by 2035.

Geothermal Power

Geothermal power involves drilling wells in areas where there is very hot rock (200–400°C) a couple of kilometres under the surface and pumping water down under pressure. The water seeps into cracks in the hot rock, heats up to far above boiling point, and comes back up another well-hole some distance away to flash into superheated steam, drive turbine blades and generate electricity. It is then captured, cooled and pumped back down again.

Countries with plenty of volcanic activity and lots of hot rock such as Iceland, Italy and New Zealand generate electricity this way – though it isn't restricted to volcanic areas. It has been several eons since the last active volcano in the United Kingdom, for example, but a 2022 report by the UK Parliament's Environmental Audit Committee estimated that the country could meet all its home heating needs and cut 20 per cent off its current greenhouse gas (GHG) emissions solely by developing its geothermal resources.[32]

> The world is really big, and the world is really hot. We've got billions of years of energy under our feet. It's all a question about how much you can access economically. We think with existing technology, drilling down to about 4,000 metres is probably cost effective.
>
> Tim Latimer, Co-founder, Fervo Energy

All geothermal activity taken together now amounts to much less than 1 per cent of non-fossil electricity generation. However, two 'fracking' techniques, hydraulic fracture and horizontal drilling, are now being exploited by a number of geothermal start-ups as a way of accessing hot, dry rock in areas that are not associated with volcanic activity. After drilling vertically down one or two kilometres, they drill out sideways and use high-pressure water to fracture the rock and create spaces where it can collect the heat. A second drill-hole at the other end of this artificial underground reservoir then delivers the water back to the surface as superheated steam, through the turbines, and then back down the first hole again in a continuous flow. The technique's competitive edge is that it is 'renewable' power that does not fluctuate, an important asset when the grid is becoming ever more dependent on wind and solar. Startups like Fervo Energy of Houston, Texas have attracted major investment, and Fervo's first megawatt-scale commercial pilot plant opened in northern Nevada in late 2023.[33]

Green, Blue and Grey Hydrogen

Hydrogen is currently being promoted to take over many of the roles that natural gas and petroleum-based fuels now fill, especially in transportation. If it secures a significant market share in energy production, this could be in the form of liquid hydrogen, but it is likelier to be as liquid ammonia, which contains hydrogen but has a higher energy density, is less flammable, and stores more easily at -33°C rather than hydrogen's -253°C. But before we get to that stage there are two key choices to be made. Will it be green hydrogen or fossil hydrogen, which is either 'blue' or 'grey'? And will it be restricted to specific purposes like steel production and fuel for aviation and ocean shipping, or will it be employed more generally as a source of energy for industrial and domestic purposes?

The most important distinction is between green hydrogen and fossil hydrogen. Green hydrogen is created by the electrolysis of water, and is carbon-neutral so long as a non-fossil fuel source generated the electricity. The electricity splits a molecule of water (H_2O) into two useful hydrogen atoms and one harmless oxygen atom. About 0.5 per cent (one two-hundredth) of hydrogen is made this way.

Blue hydrogen is made from natural gas (methane). For every tonne of hydrogen produced there are ten tonnes of CO_2 – but a portion of the carbon dioxide is captured and sequestered underground. Blue hydrogen is less than half as expensive as green hydrogen, but it still only accounts for less than two per cent of the world's hydrogen production. All the rest, almost 98 per cent of the total, is grey hydrogen, which is created in the same way as the blue sort, only all the CO_2 is just dumped into the air. For obvious reasons it is cheapest of all, and therefore dominates the market.

As green hydrogen is carbon-neutral, it can be classed as 'renewable'. Grey hydrogen obviously isn't – but neither is blue hydrogen, except in the sales brochure. Its promoters often claim that blue hydrogen is a low-emission fuel, but in fact its life-cycle greenhouse gas emissions are greater than those from simply burning the same amount of natural gas.[34]

About half the world's annual production of hydrogen (95 million tonnes in 2022) is converted into ammonia for use in the nitrogen-based fertilisers without which our agriculture would only be able to feed half of the world's eight billion people. If the hydrogen is green, then so is the ammonia, but almost all of the ammonia is made from methane and is definitely grey or blue. Another quarter is used in the oil industry to lower the sulphur content of refined petroleum. Smaller amounts are used in transportation (such as fuel cells in buses), and there is much speculation that hydrogen might be the fuel that finally de-carbonises long-distance commercial aviation and shipping.

We also know that there will be a huge growth in demand for green hydrogen as the steel industry tries to de-carbonise, and we may assume that the cost of green hydrogen will gradually fall as the volume rises: the IEA reports that green hydrogen production could reach 38 million tonnes a year by 2030 if all announced projects are realised.[35] So far, so good – but this is not where the big hydrogen play is today.

The fossil fuel interests are lobbying governments to convert existing natural gas distribution networks in many countries to handle the much trickier hydrogen, whose tiny molecules, one-eighth the size of the methane molecule, will leak through even the slightest imperfection in pipes and joints. (Big leaks can also explode.) Then, once the old distribution system has been modified at public expense, they will use it to sell us allegedly 'clean' blue hydrogen and lock us into another two or three decades of dependence on thinly disguised fossil fuels for our houses, vehicles and factories. That's the plan, and it's already having some success in the European Union and the United Kingdom.

Chris Jackson, a green entrepreneur and CEO of Protium Green Solutions, accepted the chairmanship of the lobbying group UK Hydrogen and Fuel Cell Solutions as the hype about hydrogen mounted. In August 2021, however, when he realised what the game was, he resigned, saying as he left:

> Blue hydrogen is the wrong answer. At best it is an expensive distraction, and at worst a lock-in for continued fossil fuel use.[36]

Blue hydrogen is a shameless scam. The goal should be to end fossil-fuel use, not to foster another generation of it.

One last thing. There is a remote possibility that the rocks of the planet's upper crust may contain pure (i.e. 'green') natural hydrogen in usable quantities.[37] If such hypothetical deposits were large, accessible and perhaps even 'renewable' (in the sense that the chemistry in the rocks replenishes the hydrogen supply), it would transform the entire energy economy. Even if this highly improbable miracle should be confirmed tomorrow, however, it could not be exploited rapidly enough to make a serious difference in the climate threats and policy options that governments will face in 2035.

So much for the lesser sources of carbon-free energy. Now for dreary business of dealing with the emissions that are very hard to get rid of.

Steel and Cement

The traditional processes for making steel and cement get their own special section because the associated CO_2 emissions are so large, and cannot be abated by gradual improvements. Radical changes will be required, and only now are the very first pilot plants being built that incorporate those new processes for steel. There are none yet for cement.

Portland cement, the essential ingredient in concrete, accounts for 6.5 per cent of total human emissions of carbon dioxide, and more than half of those emissions are produced during the 'calcination' process, when a mixture of limestone, clay and various proportions of iron or ash are fed into kilns and heated to about 1,450°C. Considerable energy is expended in quarrying and grinding up the materials and heating the kiln, and those inputs can be de-carbonised to some extent, but the calcination process is a chemical reaction where the outputs are calcium oxide and CO_2. No chemical reaction, no cement. No cement, no concrete. No concrete, very few large buildings, bridges and other major infrastructure.

The only durable solution to the cement problem is to create a product that has approximately the same properties as Portland cement using an entirely new process, and the search has barely begun. North Carolina-based BioMason is using sand and billions of bacteria to grow a similar structural cement at ambient temperature (the process resembles growing coral, but only takes four days). Sublime Systems in Boston makes a product identical to Portland cement at room temperature using an electrochemical process (no kilns or carbon involved). Montreal-based Carbicrete uses waste slag from steel-making, which is a limited and dwindling resource. California-based Brimstone uses basalt instead of limestone (no carbon) but still uses kilns. And none of them except

CarbiCrete is producing more than 100 tonnes a year of cement yet. The world consumes 4.1 billion tonnes of cement a year, so even in the best case it will be some decades before the industry is transformed.

The situation with steel is a little better, but the difficulty is similar: producing CO_2 is not an incidental by-product of the steel-making process: it *is* the process. Iron ore typically comes as ferric oxide ('rust'), and the iron is freed from the oxygen by heating it in a blast furnace together with a little bit of coke, which is a derivative of coal. The oxygen is seduced out of its union with the iron by the carbon in the coke and sails off as CO_2, leaving pure iron behind. The carbon dioxide released by making 2.1 billion tonnes of steel a year accounts for about 7 per cent of total human emissions.

There is an obvious answer: combine the ferric oxide ore with hydrogen instead of carbon, and it will produce H_2O (water) instead of CO_2 – and still leave pure iron behind. Add a little carbon in the form of coking coal to the molten iron at the end, and you have 'green steel'. The details are more complicated than that, of course, but if 'green steel' using this Direct Reduced Iron (DRI) process became the standard it would cut the steel industry's emissions by 90 per cent, and our global emissions by nearly 6 per cent.

Unfortunately 'green' hydrogen, produced by splitting H_2O molecules using electricity (electrolysis) from a non-fossil source, is quite expensive.

The process works fine at the laboratory scale, and in 2026 Hybrit, a subsidiary of the Swedish state-owned utility Vattenfall, will begin producing 1.3 million tonnes of 'green' steel at a full-scale demonstration plant in Gällivare, in northern Sweden. Another green steel venture, H2 Green Steel, is planning to build a fossil-fuel-free steel plant, also in the north of

Sweden, including a sustainable hydrogen facility, with production starting in 2024. ArcelorMittal, the world's second-largest steel-maker, is planning to open a commercial-scale green steel plant in Hamburg in 2025 (the German government is paying half of the 110-million euro cost), and Thyssenkrupp in Duisburg will not be far behind. It's probably safe to say that five or ten million of the world's annual two billion tonnes will be green steel by 2030.[38]

The key question is: when will the remaining 600 or 700 existing large steel plants be rebuilt to use the new technology? Steel is such a cheap product that even if the green sort cost twice as much to produce it would only raise the price of the average new car by around 2 per cent. However, retooling steel plants to use the DRI process is not cheap, and producing enough green hydrogen through hydrolysis to make one tonne of green steel currently requires 2,500 kilowatt-hours of electricity. With the existing technologies, to make all the world's steel production 'green' would require almost three times the entire global output of nuclear and renewable power as well as the construction of hundreds of massive electrolysis facilities. We know that better, much cheaper technologies for producing green hydrogen are on the way, but we cannot know when they will be available at scale.

CFCs and other Synthetic Chemicals

Emissions of chlorofluorocarbons (CFCs), hydrochlorofluorocarbons (HCFCs) and hydrofluorocarbons (HFCs) not only attack the ozone layer, they account for 3 per cent of human warming impact on the atmosphere. CFCs were banned by the 1987 Montreal Protocol, but their manufacture was allowed to continue until 1996, and all pre-1996 equipment (air-conditioning, fire suppression, refrigeration) is allowed to keep running until it dies and the CFCs leak out of it. HCFCs and HFCs – which do similar damage to the ozone layer – were banned between 2004 and 2010, but again existing equipment is exempt.[39]

As a measure of how much of these gases remains in existing equipment worldwide, the US state of Minnesota, with a population of 5.6 million, believes that its estimated 12 million residential appliances and car air conditioners contain a total of about 13,000 tonnes of CFCs, HCFCs and HFCs.[40] Extrapolating from that, let's say 750,000 tonnes is still in circulation in the United States, and 2 or 3 million tonnes for the world.

Come on, people! You started working on this in 1987, and this is how far you've got? There's fifty supertankers' worth of ozone-destroying refrigerant gases out there waiting to leak out of decrepit equipment, keep the ozone hole open and warm the atmosphere. Create an international fund to replace old pre-2010 fridges and air conditioners with new ones for free, collect the old ones, drain them safely and dispose of them. You'll never get more bang for your green buck.

You should also start working on a ban on perfluorocarbon gases (PFCs) and sulphur hexachloride (SF6), very potent greenhouse gases with exceptionally long stay-times in the atmosphere (1,000 years-plus), along with a nasty new hybrid of the two called Trifluoromethylsulfur pentafluoride (CF3SF5). Various industrial enterprises will plead that no acceptable or affordable alternative exists, but, with the possible exception of parts of the electronics manufacturing industry, the world could do without them and would suffer no huge inconvenience.

While we're at it, we should also ban dichloromethane and other new short-lived, chlorine-based gases, at least in countries that experience monsoons. Powerful monsoon updraughts are carrying these chemicals into the stratosphere, where they facilitate the destruction of ozone and help re-open the ozone hole. Get serious about this, and we could cut the 3 per cent of warming due to synthetic chemicals in half by 2030.

Oil

Oil accounts for around 40 per cent of total fossil fuel use, and nine tenths of it is burned to fuel various forms of transportation. That's what makes it so hard to ditch. Electricity from non-fossil sources can replace natural gas and coal for most purposes, but it struggles to displace oil, because the internal combustion engine requires an energy-dense liquid fuel. Nor can there be a single solution, because the engines and fuel requirements are different for ground, marine and air transport.

More than half the oil we use is burned in the world's billion and a half motor vehicles, and that is the easiest part to replace. Norway will impose a ban on sales of new petroleum-fueled cars in 2025, and at least twenty-two other countries, including all European Union members, the UK, Canada, China and Japan follow suit by 2030 or 2035.*

However, electric cars are not entirely emission-free.

Manufacturing a battery electric vehicle (BEV) involves more greenhouse gas emissions than a car of comparable size with an internal combustion engine (ICE), almost entirely due to the emissions cost of making the lithium-ion battery, but the BEV catches up after 25,000km of driving *so long as it is charged with electricity generated from a renewable source*. (Otherwise, the break-even point is much further away.)[41] However, the

---

* The United Kingdom's target was 2030 until September, 2023, but Prime Minister Rishi Sunak, seeking last-minute electoral salvation, then pushed the deadline back to 2035 as part of a panic-stricken shredding of the country's green commitments. Interestingly, British car manufacturers, who had invested heavily in electric vehicles in preparation for the change, were outraged by his betrayal. The opposition parties have promised to reverse Sunak's decision if they come to power after the next election.

real savings in emissions must await the advent of genuinely self-driving cars. Once they arrive, the transition from a 1.5-billion car world to a 300-million car world may take less than two decades.

It took New York City twenty years – from 1900 to 1920 – to go from streets full of horse-drawn carts and carriages to streets full of cars, but that was a complete change of technology. The transition from 1.5 billion privately owned cars that are parked for 95 per cent of the time to a few hundred million self-driving cars, all EVs that are publicly or commercially owned, instantly available on call on the Uber model but much cheaper (there's no driver to pay), and continuously in use, is much less demanding. Indeed, it sells itself, because it provides greater convenience at lower cost. Privately owned vehicles will still be necessary in sparsely populated rural areas, but in cities they will be rare.

It should also be remembered that there's nothing wrong with the internal-combustion engine – except that it burns fossil fuel. After 150 years of development, it's a remarkably reliable machine, and many people are working to devise synthetic 'carbon-neutral' hydrocarbon fuels that work just like petroleum products but without the emissions. (See below under 'Aviation' for details.) If these become available soon, they would speed the transition to an all fossil-free world vehicle fleet. Indeed, they might even out-compete electric vehicles, since they would not require big, heavy, high-cost batteries and would not have the same range limits. They would certainly be a better solution than hydrogen for powering heavy commercial vehicles (for which electrification is impractical because the batteries needed are just too big and heavy). However, while the chemistry for making such carbon-neutral fuels works just fine, the costs are still much too high.

There's a kind of race between electrification and carbon-neutral hydrocarbons for market dominance, and electrification has a big head start – but the consumer and the climate would win either way. And everybody wins when self-driving vehicles, long promised and long delayed, finally reach the market. The change could even come soon enough to spare us a couple of percentage points on global emissions by the early 2030s: the International Energy Agency predicts that more than a third of new car sales will be electric by 2030. On the other hand, the IEA also predicts that global oil demand will decline by only 5 per cent by 2030, because the soaring popularity of Sports Utility Vehicles (currently 40 per cent of new car sales worldwide are SUVs) is cancelling out almost all the emissions saved by the switch to electric. Global vehicle emissions will probably remain high at least until the end of this decade.[42]

Ships

Electrification is impractical for commercial shipping due to the lack of very long extension cords. This is regrettable, since around 90 per cent of international trade goes by sea, and global shipping is responsible for about 3 per cent of human greenhouse gas emissions, half again as much as aviation. But it will be a lot quicker to fix, because large ships have ample space to carry the large amounts of green ammonia (derived from green hydrogen but more easily stored and handled) that are needed to propel vessels of up to 200,000 tonnes halfway around the world.

The main reason that this switch is not already happening is the high cost of green hydrogen and the lesser but still very large cost of converting the engines of 60,000 large ships to burn green ammonia. High-volume production will bring the price down in the longer run, but green ammonia is unlikely ever to be as cheap as the marine fuel oil that most big ships use today. We should not expect a rapid de-carbonisation of marine fuel – and we might discover that we don't even want it.

The International Maritime Organisation's compulsory reduction in the sulphate emissions from marine fuel oils from 3.5 per cent to only 0.5 per cent in 2020, while cutting air pollution and avoiding an estimated 200,000 premature deaths in the next five years alone, also greatly diminished the cooling effect of those sulphate emissions. Many scientists link those cuts to the warming of one full degree Celsius in the sea surface temperature of the northern hemisphere in 2020–23 alone. Lower sulphur fuel also emits more black carbon, a warming agent so effective in melting snow and ice that there are already calls to ban these low-sulphur fuels in ships traversing the Arctic.[43] Banning fuel oil and moving exclusively to ammonia would eliminate sulphate emissions entirely and presumably cause even further warming of the sea surface. A wholesale move from fuel oil to green ammonia in marine engines might also render impractical the proposed Iron Salt Aerosols (ISAs) method of reducing methane generation in the marine boundary layer (see Chapter 7).

Everything truly is connected to everything else.

Planes

This is the hardest one: batteries weigh too much, ammonia takes up too much space and the extension cables are still too short, but burning kerosene as aviation fuel accounts for about 2 per cent of global greenhouse gas emissions. David Keith has thought long and hard about this problem.

> Let's say you are Richard Branson and you run an airline, and you're worried you might be shut down because of constraints on CO_2 emissions. You know that you can't make electric or

hydrogen-powered airplanes, and you think biofuels aren't going to work well either. One of the things you might want to do is to sell truly carbon-neutral plane seats. Direct Air Capture is a way to do that. Either by still burning petroleum in your plane, but then sucking out an exactly equal amount of carbon dioxide and putting it underground in a certified way. Or by sucking that CO_2 out of the air and turning it into a fuel by adding energy – there's no free lunch here – and that might be far more effective than simply burning hydrogen.

Let's say you have a source of hydrogen which doesn't create CO_2 emissions, and you propose to use that to solve the global transportation problem. Forget where the hydrogen came from – could be solar power or nuclear power. One option is you put it into hydrogen tanks and run fuel cells. There turns out to be immense problems with this for aircraft, because hydrogen is a very low-density fuel...

Another quite different option is you take the hydrogen, you capture CO_2 out of the air, mix the hydrogen and the CO_2, which quite easily makes a fuel – you can make octane if you want to – and then you sell a carbon-neutral hydrocarbon fuel. This is compatible with the existing vehicle fleet of the world, but the carbon in the fuel – which is octane, just like gasoline – you took from the air... They're burning it, putting carbon into the air, but then you're recapturing the same amount of carbon and selling it to them again. That's a business model that could conceivably take a whack at the global transportation market, which is the hardest part of the climate problem to crack.

David Keith

That interview took place in 2008, but Keith is still in the game. He's now the world's leading entrepreneur in Direct Air

Capture (DAC), and his 'carbon-neutral hydrocarbon' is the most promising long-term solution for commercial aviation (and many other applications as well). For the short term, however, both components of that carbon-neutral fuel, known in the industry as 'e-kerosene' – made from the hydrogen (green, of course) and the carbon dioxide delivered by DAC – are far too expensive to provide an economically viable alternative to conventional aviation fuel: four times as expensive, according to a study by E4Tech of London and Lausanne. Moreover, the first full-scale DAC pilot plants are only now being built: Keith's Canada-based Carbon Engineering company, in partnership with 1Point5, is expecting to commission its first one-megatonne-a-year DAC facility in the Permian Basin in Texas in 2024, with a similar one to follow in north-eastern Scotland. His first 'air-to-fuels' (e-kerosene) pilot plant, scheduled to open in 2025 near Merritt, British Columbia, would produce about 100 million litres of fuel annually – around eight hours' worth of global consumption of aviation fuel. So we're not likely to see the shift into high-volume, relatively low-cost production until the end of this decade at the earliest.[44]

In the meantime, we're left with Sustainable Aviation Fuel (SAF), made from renewable waste and residue raw materials such as used cooking oil and animal fat. Over its life cycle, SAF reduces greenhouse gas emissions by up to 80 per cent compared to fossil jet fuel, and it's compatible with jet engines. Production tripled to at least 300 million litres in 2022, and it is on an exponential growth path that may see it at 30 billion litres by 2030.

Currently, aircraft are only allowed to use 50 per cent SAF in their fuel, but approval to fly with 100 per cent SAF is expected in the relatively near future. The International Air Travel Association (IATA) forecasts an annual production capacity of 450 billion litres by 2050, when SAF would account for 65 per cent of all aviation fuel (with the rest presumably supplied by e-kerosene).[45]

Unfortunately, the IATA is being overly optimistic about the availability of feedstock for SAF. A recent study by the International Council on Clean Transportation estimated that SAF from waste oils and fats would struggle to supply 2 per cent of aviation fuel demand by 2030, and going beyond that level would require developing other kinds of feedstocks 'with more challenging economics and uncertain production timelines'.[46] Aviation will remain a problematic source of emissions well into the 2030s, and the ultimate solution will more likely be based on e-kerosene made by combining green hydrogen with CO_2 acquired by Direct Air Capture (though it will never be as cheap as current jet fuel).

* * *

Aviation's other big climate-related problem has nothing to do with emissions. About half of the warming caused by aircraft is due to the 'contrails' (condensation trails – aka vapour trails) that they often leave behind while flying above 25,000 feet, especially if the air is moist and cold (below -40°C). Water from the engine exhausts freezes and spreads into cirrus-like clouds that have the net effect of reflecting heat back to the Earth's surface. They are a far less serious problem than CO_2 emissions, which can linger in the atmosphere for centuries. When flying stopped over the United States for several days after the 9/11 attacks, the contrails vanished in hours and so did the warming effect. But while planes are flying, the contrails produced by the current level of global traffic are causing a transient warming that raises temperatures by up to 1.8°C at night in heavily travelled areas. (The effect is much less in the daylight hours, when the contrails also act to reduce ground temperatures by blocking some incoming sunlight.)

The simplest way to reduce contrails is to avoid areas where they are likely to form: just fly around those areas or, more commonly, below them. This would entail a modest increase in fuel usage and flight time, but the direct costs are certainly bearable: a recent study found that contrails could be reduced by more than a third just by flying at lower altitudes where the air is warmer, for an extra fuel burn of only 0.23 per cent. Alternatively, the soot that enables the moisture from the engines to condense into contrails could be reduced by 90 per cent by burning biofuels as aviation fuel instead of kerosene – even a 50:50 mix of kerosene and biofuel would halve the soot – while battery-electrical power (short flights only, for the moment) would eliminate contrails entirely.[47]

The passengers may not like the bumps at that lower altitude, but faster programmes are available that can smooth the ride out. All newer production models of Boeing and Airbus aircraft, for instance, already have 'Fly-by-Wire' installed. The aviation industry does not require massive technological breakthroughs to stop adding to our climate difficulties. It will make the necessary investments when the public and political pressure is great enough.

Fusion Power

'Fusion power is thirty years away – and always will be,' goes the old joke. Not at all, said Andrew Holland, CEO of the Fusion Industry Association, at a White House reception for the half-dozen leading American and Canadian fusion energy start-ups in 2022:

> The 2030s will be the decade of broad fusion deployment around the world.[48]

But he would say that, wouldn't he? All those start-ups are driven by venture capital, and the investors expect a return in much less than thirty years. Five to ten years is their horizon, and unsurprisingly most of the start-ups are promising 'ignition' within this decade. We are, however, not obliged to believe them.

There are occasional advances, such as a laser-initiated experiment at the National Ignition Facility (NIF) in California in 2022 that achieved 'net energy gain': the break-even point at which the experiment puts out more energy than was put into it. However, the energy release lasted only a tiny fraction of a second and put out barely enough energy to boil a kettle. Indeed, the experiment was not even designed to lead to a practical fusion-based source of energy. The NIF's main job is to help nuclear-weapons scientists study the intense heat and pressures inside explosions, which requires the lab to produce high-yield fusion reactions – but not as a reliable power source.

The much larger 35-nation International Thermonuclear Experimental Reactor (ITER) being built in southern France is intended to be the last step before the construction of a 'demonstration' fusion power plant that could be hooked up to the grid – but the 'demonstrator' is scheduled for some time in the 2040s. The sheer diversity of approaches among the private companies that have now entered the nuclear fusion field might produce a breakthrough sooner than that, but at the moment commercial-scale fusion power is still at least thirty years away, making it irrelevant not only to 2035, but also to 'Net Zero by 2050'.

* * *

Here ends our survey of the options for de-carbonising our energy system. The picture isn't as bad as it looked even five years ago. True, most of the options for cutting emissions will not yield significant cuts before 2030 at the earliest, but solar and wind power are doing very well indeed. We will draw more benefit from the electrification of the world's vehicle fleets than the deadline dates suggest, because sales of new petroleum-fuelled cars will probably collapse a good five years before they are banned. A pivot back to nuclear, which already seems to be happening in the Global South, would also help if it comes early enough. And big cuts in methane emissions (up to 30 per cent in a decade) would be much easier to achieve than equivalent cuts in carbon dioxide.

So if you have hopes of avoiding +2°C, you will be pleased to know that we are still in the game.

* * *

1.5°C is already bad but 3°C is potentially civilisation-ending bad. In between is where we're rolling the dice.[49]
Prof. Michael E. Mann, Director of the Center for Science, Sustainability & the Media, University of Pennsylvania

In October 2021, the journal *Nature* contacted all 233 authors of the report on climate change that went to the Intergovernmental Panel on Climate Change (IPCC) and provided the scientific basis for the negotiations at the COP26 summit that year in Glasgow. Ninety-six scientists – 40 per cent of the group – replied to *Nature*'s questions asking their private opinions about the probable course of the warming. Six out of ten said that they expected the world will warm by at least 3°C by the end of the century. More than 80 per cent said they expected to see catastrophic impacts of climate change in their own lifetimes (therefore, by 2050 or 2060, in most cases). Only 4 per cent believed that the world might be able to limit the warming to less than +1.5°C. So if there were some sort of climate insurance available against 'catastrophic impacts', I'd be inclined to buy it.

The only insurance of that sort is called 'geoengineering' or 'climate engineering', and it's a safe bet that ten years ago 90 per cent of the scientists in the *Nature* poll would have regarded the very idea with horror. My unscientific estimate, based on almost a hundred recent interviews, is that about half of them would now be willing to consider it as a last-ditch temporary measure to hold the heating down and win us more time to de-carbonise. Desperate times call for desperate measures.

> It is a very dark point we have reached... The scientific community has transitioned over just the last ten years from being very sceptical of even debating geoengineering, and certainly not considering it as part of a solution, to now having it being forced into the main room of the solution space. Because when you look clearly at our options... the remaining carbon budget is something like 300 gigatonnes of CO_2 and we are emitting 40 gigatonnes per year. You just divide 300 by 40 and you see that we have less than eight years left under current levels of emissions. It means that we don't only need all hands on deck when it comes to de-carbonising; we also need to truly consider all options.
>
> Johan Rockström, Director of the Potsdam Institute
> for Climate Impact Research

That was in 2020, so four years left now.

7

LAND AND FOOD

Three-quarters of our greenhouse gas emissions come from burning fossil fuels; almost all the rest come from working the land. If you find the strength of vested interests and the fear of 'stranded assets' frustrating in the field of fossil fuels, their power in the domain of land and food will drive you crazy. There is no older vested interest than land ownership, and farmers don't just own what may become stranded assets; they and their families usually live on them.

Humanity has managed most of the planet's land one way or another for thousands of years, but our huge numbers and our impact on the Earth System now require us to make several urgent and difficult decisions. We have to reduce the greenhouse gas emissions from our agricultural activities as part of the larger task of stopping the warming. Equally important for that goal is restoring as much land as possible to its original biogeophysical role as the largest carbon sink on the planet. That means taking it out of agriculture and 'rewilding' it, which is necessary anyway to reverse the accelerating collapse of biodiversity.

Given that we removed this land from its natural functions in order to produce food for human beings, however, the amount that we can release back to natural ecosystem services, and how quickly, will obviously be constrained by the need to go on feeding our population. It will also be limited by our ability to find new homes and roles for the farmers who lose their land and jobs as a result of this process.

Not just a tall order: a whole forest of very tall orders. But we have to start somewhere, and we might as well start with the easiest part. There is some low-hanging fruit that might let us halve our greenhouse gas emissions from farming by 2035 – which is pretty much the minimum requirement right across the board if we are going to keep the rise in the average global temperature later in the century below +2°C. The fruit in question goes 'Moo'.

> When I do a puzzle with my daughters, there is usually an elephant next to a giraffe next to a rhino. But if I was trying to give them a more realistic sense of the world, it would be a cow next to a cow next to a cow and then a chicken.[50]
> **Professor Ron Milo, Weizmann Institute of Science**

Human beings, by weight, account for 36 per cent of the mass of mammalian life on the land surface of the planet. Wild mammals account for only 4 per cent. All the rest, amounting to 60 per cent of the whole, are livestock bred, raised, slaughtered and eaten by human beings – and we control the future of our livestock.

If we were to find ways of replacing the services that they provide us with their meat, milk and eggs (as well as cheese, butter, leather and wool), we could cut their numbers drastically and free up a large proportion of the Earth's surface for growing plant food directly for human beings, or for reforesting and 'rewilding' to restore some of the natural systems we have ravaged.

Nobody would have to go naked and barefoot if we stopped using wool and leather; adequate substitutes are already available and inexpensive. Replacing meat, milk and eggs wholesale has been seen until recently as a very difficult and perhaps impossible task, but that situation is beginning to change.

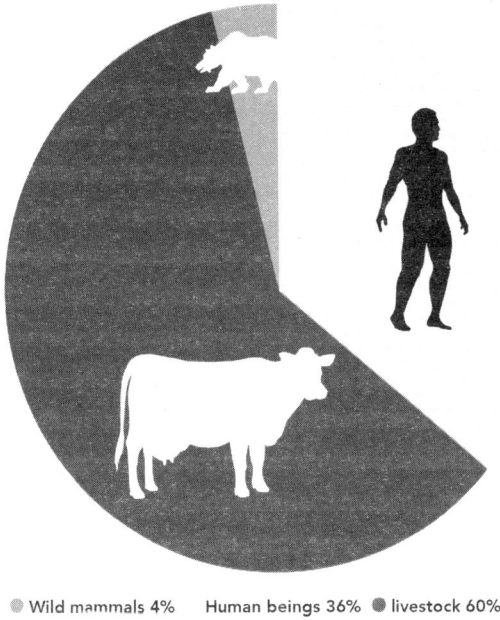

● Wild mammals 4%   Human beings 36%   ● livestock 60%

Mammalian life on Earth by mass

Meat Alternatives

> But people have always eaten people.
> What else is there to eat?
> If the Juju had meant us not to eat people
> He wouldn't have made us of meat.
>
> Song: 'The Reluctant Cannibal', Michael Flanders and
> Donald Swann (1956)[51]

There is, to begin with, a steep increase in the number of people following vegetarian or vegan diets, although their food still must be grown by traditional agricultural means. Then there are the new 'cellular' or 'clean' meat products, grown on lattices from stem cells harvested from living animals, that could satisfy the needs of those who won't give up the traditional diet but are willing to pay a good deal more for environmentally harmless and morally irreproachable meat. And a little further off, but already in prototype versions in Finland, Israel and the United States, are proteins grown directly from micro-organisms, hydrogen and carbon dioxide that can be synthesised into foods of all kinds for animals and people alike – 'food from the air' or 'farm-free food'.

The fundamental issue is land use. Humankind has appropriated 40 per cent of the land surface of the planet for its agriculture (up from 7 per cent in 1700), removing both the trees and most of the original wildlife and replacing them with crops and food animals. (If you count 'managed' forests, roads, buildings, ski-runs and everything in between, we control 75 per cent of the ice-free land surface.) We have increased the mass of animal life on the land fourfold, but removed two-thirds of the mass of vegetable matter and set most large animal species we don't make use of on the road to extinction.

This directly affects the welfare and ultimately the survival prospects of human beings, because by cutting down approximately half (46 per cent) of the original forest cover in the past 12,000 years, we have halved the effectiveness of the largest carbon sinks on land. In planting food crops and other commercial plants like tobacco and cotton on the arable land and repopulating the new grasslands with a far larger population of domesticated large animals than the wild animals that once lived in the lost forests, we have created sources of methane and carbon dioxide emissions that account for 18.4 per cent of all human greenhouse gas emissions. Explanations about exactly where these emissions come from vary according to the perspective of the explainer (*Cowspiracy* and so on), so we should break that number down by specific sources.

Agricultural Emissions

The biggest single source of greenhouse gas emissions in agriculture is livestock and manure: 5.8 per cent of the total.[52] Most of this is methane produced by ruminants: cattle, sheep and goats, all of which burp or fart large amounts of methane into the air as they digest their food. Pigs, chickens, and other farm animals also contribute their bodily wastes to the problem, but their digestive systems are different: more than 90 per cent of the methane in the mix comes from cows. (Methane is a particularly powerful greenhouse gas, but for purposes of comparison the total impact on the atmosphere of all the gases emitted by livestock is expressed here in terms of an equivalent amount of carbon dioxide.)

Given the large proportion of crop-growing that is devoted to animal feed, we should also count half of the total emissions cost of nitrogen fertilisers to animal production, and more than half the nitrous oxide and methane from decomposing organic matter in wastewater as well. That gives 5.8 per cent+2.1 per cent+0.1 per cent = 8.0 per cent as the real emissions cost of producing meat. It's the largest single source of emissions after fossil-fuel burning.

The next biggest source of agricultural emissions is crop-burning (more precisely stubble-burning or waste-burning), which accounts for 3.5 per cent of global greenhouse gas emissions. Farmers often burn crop residues after harvest to prepare land for the next planting, and this may be an easy target. Changing old habits is never easy, but it should be possible to devise less harmful ways of preparing the land.

Beyond that the targets get smaller: tree-cutting, tree-planting and changes to carbon stores in forest and ex-forest soils: 2.2 per cent of net emissions. Nitrogen fertiliser and other soil treatments in non-meat agricultural production: 2.0 per cent. Emissions from disturbed soil in crop-lands: 1.4 per cent. Methane emissions from water-logged rice fields: 1.3 per cent. Net carbon losses from grassland biomass and soils: 0.1 per cent. Total for direct emissions from human agriculture: 18.4 per cent.

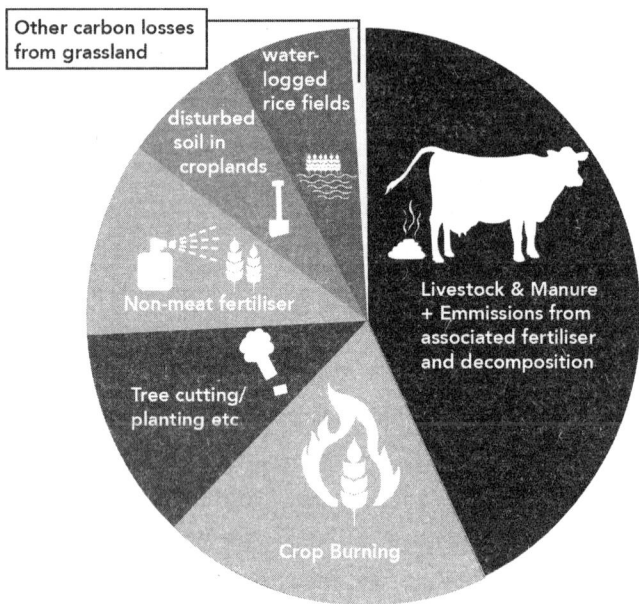

Other carbon losses from grassland

water-logged rice fields

disturbed soil in croplands

Non-meat fertiliser

Tree cutting/planting etc.

Crop Burning

Livestock & Manure + Emmissions from associated fertiliser and decomposition

Emissions from agriculture

So what could we do to reduce the various sources of food-linked emissions by 2035?

Meat

Meat is not an indispensable part of anybody's diet. Animal husbandry is directly responsible for 8 per cent of global CO_2 emissions, and it would make a real difference if those emissions could be halved by 2035.

There are at least one billion cattle in the world.* To reduce that number by half in a dozen years would certainly be ambitious, especially since India's share of the cow population is sacred, but it is technically feasible in the sense that production of non-meat foods could expand enough by 2035 to replace the lost meat. The harder question is how to persuade everybody from farmers to consumers to accept such a rapid, steep cut in the consumption of a food with deep roots in most cultures, but let's just imagine for now that we can eliminate half the cows on the planet by 2035 and achieve similar cuts in other livestock numbers from pigs to poultry. That would save half of the 8 per cent of emissions directly caused by raising meat, so a 4 per cent cut in global greenhouse gas emissions there.

Fodder, Crop Waste, No Till

Our livestock cuts would also take at least 2 per cent out of the 3.5 per cent of CO_2 emissions that are caused by crop-burning, because fewer animals means less demand for maize and soya. (Fodder for animals accounts for almost half the grain we grow, and if you didn't grow the crop, you don't need to burn the crop waste.) Assume that you can sweet-talk another 1 per cent out of the crop-burning total just by offering farmers a better way of preparing the land for the next crop (with relevant subsidies if necessary).

* * *

You are now sitting on a total 7 per cent cut in global emissions. Congratulations. After that it gets much harder. I won't go through an exercise in fantasy accounting for the smaller sources

---

* Estimates vary but the United Nations Food and Agriculture Organization says there are 1.5 billion cows in the world, in which case there are approximately equal numbers of cows and of cars. Their emissions are also approximately equal, but the cars emit mostly CO_2 whereas the cows favour methane (CH_4).

of direct agricultural emissions (such as methane from water-logged rice-fields and nitrogen fertiliser for crops for human consumption), but we'd be lucky to halve them in a dozen years, so say another 1 per cent off global emissions there. Now the total's up to 8 per cent – but that's where we have to stop.*

Not bad for a dozen years – but remember that these calculations are based on the assumption that half the world's cows and sheep will magically disappear.

Fake Meat Alternatives

How might we make them disappear? One way would be simply to order a massive cull of the cattle and sheep at the national or international level, but most of them are private property and doing that is politically unthinkable unless and until the climate crisis starts killing people in the tens of millions. Maybe not even then. The alternative is to shrink and ultimately kill the demand for the meat and milk that is the commercial reason for keeping all those ruminants. If the demand collapses, they really will magically disappear.

For the many people who enjoy the taste and texture of meat but are concerned about the greenhouse gas emissions, just eating less of it can make a significant difference. So does switching from beef and lamb to pork, poultry and/or fish: the non-ruminant food animals involve only one-fifth of the GHG emissions per kilo of beef and lamb. For those willing to go further, the fake-meat options are expanding fast in the richer parts of the world. Go to the vegan and vegetarian section in any supermarket and the shelves are already crowded with food and

---

* Agriculture also involves carbon costs not directly linked to the chemistry of food production (farm machinery and its fuel, refrigeration, food processing, packaging and transport) which amount to another 7 per cent of global emissions, but these are almost all energy costs and have been accounted for in the previous chapter.

drink made from plants that claim, with greater or less accuracy, to taste just like meat and cow's milk (or even better). Further down the road (and certainly higher in price) are the 'cultured' meats grown on lattices from beef or pork or chicken stem cells, and milk and dairy products grown by a similar process: real meat and dairy without the same emissions and pressures on the land.

We are currently a long way from seeing a steep decline in global meat consumption, but we may be approaching an inflection point. Global meat consumption is still growing – the UN's Food and Agriculture Organization (FAO) predicts that the world meat consumption will grow by 14 per cent in the 2020s – but that is now driven mostly by population growth. *Per capita* meat consumption is growing by only one third of one per cent per year. Moreover, the share of 'red meat' in the mix is slowly declining: more than two fifths of world meat consumption will be chicken by the end of this decade.[53] Yet crucially, the share of 'ruminants' that eat grass (cattle and sheep) is not falling. They only account for one-quarter of the total meat consumed (beef 20 per cent, lamb 5 per cent), but are responsible for almost two-thirds of all the greenhouse gases released during meat production.

The problem is not 'red meat', as is commonly supposed. Pork, the dominant red meat consumed in East and Southeast Asia, produces little more CO_2 per kilo than chicken or fish. (Pigs – 5.77kg; poultry – 3.65kg; fish – 3.49kg). Cattle and sheep, by contrast, have a highly specialised four-chamber stomach in which the main work of digestion is carried out by microbes that effectively ferment the grass so that the animals can absorb the nutrients. It is this fermentation process that makes the ruminants such prolific producers of greenhouse gases: 26.61kg of CO_2 per kilo of beef and 25.58kg per kilo of sheepmeat (mutton or lamb).[54]

To make matters worse, most of the greenhouse gas emitted by ruminants, although conventionally measured in 'carbon dioxide equivalent', actually comes in the form of methane (the cattle belch it out; the sheep fart it). Methane is a relatively short-lived gas – most of it will decompose into carbon dioxide within ten years – but during that decade it has eighty-six times (that's *eighty-six times*) the warming effect of a comparable volume of CO_2. So if you happen to live on a planet that is teetering on the brink of runaway warming, methane is not your friend.

This presents us with an unfamiliar perspective. From a climate-oriented point of view, the problem is not 'meat' or even just 'red meat' (like pork), and frankly even sheep, at only 5 per cent of total meat consumption, are not a very big deal. The problem is specifically cattle, in both their meat and dairy roles (around a quarter of cattle are dairy cows). These seemingly inoffensive animals are responsible for at least 7 per cent of our greenhouse gas emissions (and perhaps much more than that), and the food they consume accounts for around half of all the land we have taken for agriculture. They are therefore a very big problem.

In principle, at least, there is a solution to this problem, but first let's consider the other, related problem of how we are using the planet's land.

Restoring Wild Lands

Urgent action is needed on 'restoring the carbon sinks', a deliberately evasive phrase that disguises the nature of that task. We are dumping around 40 billion tonnes of CO_2 into the air each year. About a third of it is absorbed by the oceans, but far less is soaked up by the land surface of the planet because we have cut half the Earth's forests down. The lost 'carbon sinks' are really the missing forests, and 'restoring' them means putting the forests back. Not all of them, obviously, because we're using that land to grow our food on, but as much as possible because we're also losing the race to hold the temperature down. Even the Intergovernmental Panel for Climate Change now admits that if we don't restore the carbon sinks we have no hope of staying below +2°C.

How much is possible? More than you might think, because almost half of our farmland is devoted to growing animal feed, mostly for our cattle. If people eat less beef, much of that land is taken out of production, and (in theory, at least) the land becomes available to 'rewild'. There is no government programme anywhere that is designed to subsidise this process today, but if the magic somehow happened and the cows started to disappear, huge amounts of our existing farmland would become available for recreating the lost carbon sinks on land.

Realistically, weening humans off beef will take at least two or three decades (i.e. about the same length of time it took Western countries to reduce smoking to a fringe activity). To avoid pushing both the climate and biodiversity off a cliff in the meantime, therefore, we need two apparent miracles: one to stop more than a billion cows burping methane, and another to feed them in a way that allows us to give back much of our agricultural land to restore the carbon sinks. Pilot programmes are already underway on both fronts.

Carbon Sinks and Precision Fermentation

Solar Foods is in Espoo, a distant suburb about half an hour's drive west of Helsinki. It's housed in a row of rented garages in a street just off the highway, with no name on the building because they have already attracted attention from various zealots. Co-founders Pasi Vainikka and Juha-Pekka Pitkänen started with a micro-organism they found 'in a Finnish bog somewhere' (it's already patented), grew it in a modest amount of water in a bioreactor, drained off the soup of single-celled creatures (300 grams a day) and dried it. It ended up as a bright yellow powder with the consistency of flour that is 65 per cent protein and ready to be processed into any kind of food you want. They call it 'Solein'.

> This is the first time in history humankind can be provided with edible calories that at no point require photosynthesis. So far photosynthetic plants have been the only feasible way to receive energy from the sun to feed humankind. Now, this process can be bypassed in its entirety. That's an absolutely historic moment. A new era begins in the primary production of food and restoring biodiversity.
>
> Pasi Vainikka

According to the blurb for the first Solein-based five-course 'fine dining' meal, served to the public in a Singapore restaurant called Fico in May 2023, Solein is 'a versatile protein that has an enormous number of functions. You can use it as a structure-forming ingredient to produce many different textures and tastes that would have a protein component in them in sauces, spreads, beverages, noodles or pasta, baked goods or in foods to replace meat, dairy, or eggs.' Here's the menu itself:

Kanzai-style Ozoni Misolein Soup
Solein Pasta with Singapore Pesto
Smoked Pumpkin with Solein Salted Egg Sauce
Fennel Bergamot Granita with Lime and Fennel
Solein Bean Curd
Solein Powered Ice Cream with Figs and Fig
Vincotto

Why Singapore? Because it is a prosperous island state smaller than Disney World with almost six million people on it. It must import *all* of its food and it's nervous about climate-related shocks to its food supplies, so it has a resilience strategy called '*30 by 30*' that aims to produce 30 per cent of the country's food domestically by 2030. That means producing the food in factories, not on farms, so Solein was an obvious candidate.

But while the human market for 'food from the air' will doubtless grow enormous in the long run, the biggest near-term market will be bulk protein for animal feed. You don't need to cut down the Amazon forest to grow more soybeans if you can just produce an almost identical product in bioreactors. Solar Foods will begin commercial production of Solein in its new 'Factory 01' at Vantaa (near Helsinki airport) in mid-2024. Half a dozen rivals in other countries are close behind.

Solein is not cutting-edge technology: bioreactors are old hat, as is the care and feeding of desirable micro-organisms. All they need in order to double their numbers every three hours is hydrogen (obtained by splitting water molecules with solar power), carbon dioxide from the air, water and light. The microbes get their energy from processing the hydrogen, so they don't even need sugar. No land at all is involved in the process except the hardstand that the equipment sits on. Now, for the first time, there is an emerging climate-change-driven mass market for the food products that the microbes can produce.

This is a technology that might enable the human race to return most of the agricultural land we now use for growing animal feed to its original ecosystem roles, but (being socially responsible Finns) the founders of Solar Foods are also keenly aware that there will need to be alternative jobs for all the farmers who used to grow that animal feed – and ultimately for many who grow food for human beings too.

It's a question of how we use the land. Currently we are using it to cultivate crops, but we could also cultivate forests... In Finland, at least, many farmers are already tending forests, but this could be more in the future. They wouldn't be growing wheat or tending pigs; they'd be growing and tending the forests.

Juha-Pekka Pitkänen, Co-founder, Solar Foods

The rub is that there won't be as many jobs for the tree-tenders as there were for the farmers. One new job for every ten old ones, maybe, and clearly the other nine out of ten farmers would have something to say about this scenario. Juha-Pekka Pitkänen doesn't have a good solution to the problem any more than I do. He wants his biofood start-up to succeed, and so do I. There's just a huge log-jam of political and moral problems in a technology that could put most of the world's farmers out of work.

This [Solar Food product] is 65 per cent protein, so we still need wheat, corn and rice. The human diet consists in large part of carbohydrates, and that would mean some kind of plant production, so some agriculture is still going to continue. But growing animal feed, any kind of animal farming, will go smaller and smaller.

Juha-Pekka Pitkänen

Harpoon Weed and the Cow Problem

Most of the cows probably will have to go in the end, but a shrinking cow population might be maintained for several decades without a climate apocalypse if Solar Foods-style animal feed derived from 'precision fermentation' largely replaces the field-grown cereal grains and oil meals (soybean, canola, etc.) that dominate the animal-feed market, *and* if some way is found of curbing the immense greenhouse gas emissions from cattle.

> If just 10 per cent of global ruminant producers adopted *Asparagopsis* as an additive to feed, it would have the same impact for our climate as removing 50 million cars from the world's roads.
>
> Damien O'Connor, Agriculture Minister,
> New Zealand (2019)

Ruminants are cattle, sheep and goats. *Asparagopsis armata*, also known as Harpoon Weed, is a red marine seaweed that was found in Australia almost two centuries ago, but now grows all over the Southern Hemisphere. New Zealand is a country where the cows (and for that matter, the sheep) greatly out-number the people. The methane these two animals produce accounts for almost half of New Zealand's greenhouse gas emissions. The country is legally committed to reaching net-zero emissions by 2050, so it makes sense for this to be a major focus.

Research led by New Zealand scientists showed in 2019 that adding only a 2 per cent supplement of *Asparagopsis armata* or its close relative, *Asparagopsis taxiformis* to the animals' feed can reduce the methane emissions from cattle and sheep by 82 per cent. If those results can be duplicated globally, then the damage to the climate from livestock rearing could be reduced much faster than we thought. If a small portion of that seaweed were added to all cattle and sheep feed worldwide – highly unlikely, but stay with the notion – that alone would give us our aspirational 4 per cent cut in global emissions from live-stock farming without having to make massive cuts in the total number of animals right away, which is very unlikely to happen.

Of course, there are a couple of problems with this approach.

The first is we almost certainly won't be able to produce *Asparagopsis* in the required amounts, because 2 per cent of all the animal feed consumed in a year by cattle and sheep is a gigantic volume of plant matter.

> The holy grail is going to be getting enough of the stuff. It's got a complicated life cycle. One of the barriers to getting this to market is growing enough of the stuff in ways that are cost effective to make it commercially viable.
>
> **Professor Charles Eason, Cawthron Inst., New Zealand**

This seaweed will grow in other parts of the world (it has been successfully introduced in north-western Europe and the Mediterranean), but it requires shallow coastal waters, which are relatively limited in extent and are already fulfilling many other ecosystem functions. So in addition to growing as much of it as we can, we should be looking for other plants or synthetic equivalents of *Asparagopsis* that have a similar effect on methane production in livestock.

Not quite so dramatic in effect but closer to volume production is a cattle-food additive called Bovaer that has been under trial by the Dutch-based DSM company for the past ten years. It promises average reductions in methane production of 30 per cent for dairy cattle and 45 per cent for beef cattle, with some herds scoring even higher. It has no negative effects on milk production or animal health, and involves the addition of only a quarter-teaspoonful a day to the animal's feed to suppress the enzyme that triggers methane production in the cow's *rumen* (stomach/fermentation chamber). It also works with sheep. DSM expects the price to be 50–55 euros per cow per year once its first large-scale production facility opens in Scotland in 2025. Bovaer is now authorised for sale in 45 nations including the EU countries, Australia, Brazil and Argentina, and there are no critical resource constraints on expanding production.[55]

How Rewilding could Work

What would a successful global effort to transform the world's agriculture look like by 2035? One can imagine a benign future where the rapid spread of precision fermentation products like Solein for animal feed has taken the pressure off the land and lowered its price as well. State-funded programmes soften the blow to the landowners by buying up farmland for rewilding, and the generations-long project of returning half the agricultural land to its original ecosystem roles gets underway. Priority is given to assembling very large blocks of land with little or no resident human population.

Meanwhile, meat-eating is being eroded not only by generational turnover but also by competition from 'farm-free' foods for human consumption that can mimic traditional dishes but have also diversified the menu. The ruminants' methane problem has been solved by synthetic versions of *Asparagopsis*, and the steady flow of farmers who give up are bought out by government at generous rates and assisted in finding new jobs and homes. The demand for 'real' meat is shrinking, the cow population is dwindling, and more land is coming available for rewilding. By 2060 almost half the land once farmed by people has been given back to nature, and everybody lives happily ever after.

Yes, one can imagine it, but one has great difficulty believing it. Human beings don't work like that, or at least they haven't done so very often up to now. Many mistakes will be made, and there will be delays, conflicts and victims. Maybe no substitute for *Asparagopsis* will be found, and in a world where the climate has tripped too many feedbacks and is going runaway we are forced to ban beef and live with the consequences. Maybe misguided fanatics create a wildly popular 'Back to the Land' movement and sabotage the rewilding project. There are no guarantees. By all means think of the above idyll as a plausible template for a relatively successful and pain-free transition, but not as a description of the probable future. Some of it will happen, though, and more of it could.

Attack of the Sunset Industries

The long-term solutions are visible and credible, but it is hard to believe that the entrenched resistance to such radical changes as those above can be overcome in just a dozen years. The understandable desire of farming families to preserve the status quo, combined with the determination of the agribusiness sector to protect its business model, will together represent the biggest and best-funded lobbying group in human history. Big Tobacco managed to stall action on smoking for thirty-one years simply by spending huge amounts of money on lobbying and advertising, even though smoking effectively kills about half of all smokers.* Big Oil did even better: way back in 1977 ExxonMobil's scientists accurately predicted a global temperature rise of 0.2°C per decade if fossil fuel use continued on its present trajectory, but the industry's lobbyists and PR experts – some of them the same people who delivered for Big Tobacco – succeeded in casting such public doubt on the findings of climate science that they postponed the first ban on the sale of petroleum-fueled cars in a major country for another fifty-eight years.[56] (A ban on the sale on new fossil-fuel-powered cars comes into effect in many countries in 2035.)

Unless Big Meat (or Big Aggro, if you prefer) operates to far

---

* Initial US Surgeon-General's report that smoking kills – 1964. First large-scale ban on smoking in indoor workplaces – 1995 (in California).

higher scientific and moral standards than cigarette-makers and car manufacturers, we may expect the same type of campaign to be waged by the agro-industrial corporate giants against major reductions in the global cattle herd or the area being farmed for grain. Transnationals like Tyson Foods and Cargill (head-quartered in the United States), JBS and BRF (Brazil), Vion Food Group (Netherlands) and the WH Group (China) will quickly figure out (if they haven't already) that there will be no radical reductions in the world's herd size or grain-growing areas unless the supply of 'food from the sky' grows rapidly, and this will be the terrain on which they choose to fight.

Patents will be bought up and buried, influencers will be 'encouraged' to spread fear of the new food technologies online, expensive nuisance lawsuits alleging safety issues will be brought against start-ups. All the old tricks and a few new ones will be deployed to preserve the present structure of food production and the associated cash flows.

Nor will Big Meat be alone in this struggle. Unlike Big Tobacco or Big Oil, which had few allies outside their own industry, Big Meat can count on the instinctive support of almost all farmers and their children: about two billion people, or *a quarter of the human race*. Expect to hear a lot about 'the tragedy of the farmer' and expect the process to take a long time. 'Sunset industries' typically fight long and hard to stop the sun from going down on them, even though they know that the sun must eventually set. In the end they almost always lose, but a lot of money can still be made before the end arrives, and a lot of time can be lost.

The maximum contribution of the food sector to global emissions cuts in the next dozen years is unlikely to exceed 2–3 per cent, far from the ideal 8 per cent I proposed at the start of this chapter.

Methane: Leaks and Dumps

If you really need to cut your greenhouse gas emissions fast because you are nearing tipping points, you have to prioritise the short-term over the long haul – you won't reach the long haul if you don't survive the short-term crisis. Methane accounts for at least a quarter of anthropogenic warming, and it's a promising target for quick cuts because it stays in the atmosphere for only about a decade, as opposed to several centuries for CO_2. Excess carbon dioxide is the biggest long-term threat because it builds up over centuries, but concentrating on cutting methane emissions is better if you're in a jam. Make big cuts in your methane emissions now, and you'll be seeing a real effect by 2030.

> We can now detect methane leaks very easily. A wonderful student of mine has mapped all of London's major streets and their gas leaks. We have an instrument called a cavity ring-down system that measures methane. She basically has a sniffer up on the roof and drives along the road, detecting leaks. Just like a fox can smell a chicken coop at two miles.
>
> Euan Nisbet, Department of Earth Sciences, Royal Holloway, University of London

The cuts in methane can come fast because in 2022 the Global Methane Pledge committed 105 countries to cut their emissions by 30 per cent by 2030. The pledge didn't include China, India, Russia or Australia – all major emitters – but it's still a big deal. Its primary target is the methane leakage from oil and gas infrastructure, which in 2020 amounted to 70 million tonnes, mostly from flaring the methane and from the world's 750,000km of natural gas pipelines. Most of this can now be located cheaply and easily by 'sniffers' in town, and by satellites and/or drones in the country. After that, it depends on local law enforcement (and repairing the leaks can save the producers money).

Another easy target is dumps and landfills, especially in the tropics where they are poorly regulated and release large amounts of methane as their contents rot. All you have to do, as Euan Nisbet explained, is to shovel half a metre or so of soil on top 'because the soil bacteria will say "Oh methane. That's food. Let's eat that." It's dead easy to cut the landfill emissions.'

An even bigger opportunity is food waste. About one third of all the food that is produced – 1.3 billion tonnes – is wasted. So is the 30 per cent or so of agricultural land on which that food is produced. There is ample room for improvement quickly and at low or no cost.

The amount of food wastage varies across the categories: 45 per cent of fruit and vegetables, quick to spoil and with a low cost per kilogram, but only 20 per cent for milk and meat, also quick to spoil but with a high value. It is also very sensitive to location: in developed countries consumers and retailers throw away between 30 per cent and 40 per cent of all food purchased, whereas in poorer countries only 5–16 per cent of food is thrown away. The big losses in the developing countries happen due to inadequate infrastructure: poor storage, refrigeration and transportation. So the remedies must also be different.

In the richer countries, it's the consumers who must change, and they can if they are prodded. The UK Government-backed 'Love Food Hate Waste' campaign reduced food waste in the United Kingdom by 12 per cent – more than a million tonnes a year – between 2007 and 2012, and the amount lost continues to decline at a slower rate today.[57] In poor countries, by contrast, consumers waste almost no food. Most of the wastage occurs between the fields and the retailers, and what is needed is investment in infrastructure for the food industry.

The wastage can never be eliminated, but it is possible to cut it globally by half in a dozen years. Consumers clearly can change their habits (many of the current generation of teenagers would be happy to serve as the Food Waste Police), and the capital costs involved in upgrading infrastructure are modest given the benefits in terms of reducing greenhouse gas emissions (most wasted food ends up rotting and emitting methane). Moreover, ending the need to grow food on about 700 million hectares of land would be a major achievement.

And finally, we are beginning to see proposals for 'Methane Removal' comparable to the much more advanced projects for Carbon Dioxide Removal. The logic is the same: if you didn't manage to prevent the emissions, you can also try to chase them down and reverse them. All the proposed techniques are embryonic – you can see a comprehensive list at *Spark Climate* – but the one getting the most attention at the moment is Iron Salt Aerosols (ISA).[58]

> There's a lot of naturally occurring sea spray aerosol that contains salt. There's also a lot of naturally occurring mineral dust in the atmosphere that contains iron. But when the iron is combined with the sea spray aerosol, something rather unique happens because you form iron chloride, which absorbs sunlight and creates chlorine atoms.
>
> **Professor Matthew Johnson, Atmospheric Research Centre, University of Copenhagen**

Matthew Johnson is looking into a well-known natural process where iron-rich dust blown over the ocean interacts with sea-salt particles (sodium chloride) to form iron chloride molecules. Those are stable in the dark, but in sunlight the iron chloride molecules tend to release one of their chlorine atoms as a free radical (missing one electron). And these chlorine radicals are very efficient destroyers of methane.

As things stand, this process doesn't remove much methane from the atmosphere (only about 3 per cent of total global methane removal), because there isn't much iron-rich dust over the oceans to form iron-chloride molecules and kick-start the process. Johnson is researching what might be accomplished if more iron were made available, perhaps as an additive to the fuel of ocean-going merchant ships.

> If you look at the price of the different methods of reducing the amount of methane in the atmosphere, some things are very low cost – plugging natural gas leaks, for example. But you quickly run out of those inexpensive costs, and when you get to somewhere around 30 per cent, the price just skyrockets. It's at that point that removing methane from the atmosphere [by destroying it with chlorine] becomes the cheaper alternative.
>
> Matthew Johnson

It's still not easy. Destroying methane is so expensive because it's 200 times less concentrated in the atmosphere than carbon dioxide, which means only passive methods are affordable (no giant fans moving air around as in Direct Air Capture for carbon dioxide). Moreover, methane is hard to work with because, as Johnson put it, it's 'nearly a noble gas in its chemical structure'. In other words it doesn't stick to surfaces, and it's very hard to get it to adsorb on to anything.

> That's why this is such an exciting breakthrough: we found the Achilles heel for methane, and that's the reaction with chlorine. By introducing iron, you create a super aerosol that's a far better catalyst than what we have going on

today. One iron atom when introduced could produce tens
of thousands of chlorine atoms, depending on the catalytic
efficiency.

Matthew Johnson

The objective is to halve the amount of methane in the
atmosphere, all the way back to the pre-industrial level. The
method is plausible, it's affordable – just add the iron dust to
the ship's fuel exhaust and send it up the funnel – and it could
be deployed in less than a decade if no deal-killers show up on
further investigation. We could use a miracle around now.

I realise that the measures suggested above are a lame
second-best to the programme for restoring half the
agricultural land to natural ecosystem services that we should
ideally be pursuing, but I also know the issue of rewilding the
land is the rock on which the whole enterprise could founder.
Not making it an absolute priority will have grave costs in
lost biodiversity, but the conflicts caused by a full-scale 'land
reform' project that puts most farmers out of business would
be a killer distraction from the primary job of holding the
average global temperature down. There may be a few rich
countries that can start buying their farmers out, and more
power to them, but in most places even to attempt such a
thing now would cause at least political paralysis, if not actual
rebellion.

National programmes for returning half the farmland to
nature, agreed by consensus, supervised by government, and
with fair compensation for all who have to leave the land, would
be wonderful, but we are not that species. We may manage to
do it, in the end, but not without huge conflict, great delay and
some bloodshed.

Meanwhile, a gradually reducing cow population might be

maintained for a generation without a climate apocalypse on two conditions: that animal feed derived from 'precision fermentation' replaces the field-grown cereal grains and oil meals (such as soybean and canola) that dominate the animal-feed market, and that a way is found to curb the immense greenhouse gas emissions from cattle. Pigs, poultry and farmed fish could remain on the menu for a long time, because they cause so much less harm to the atmosphere than beef and lamb meat.

As for what to do with the land that is released from agricultural use, reforestation is often the best bet, but not by commercial monocultures, and not 'afforestation' – planting a forest where there wasn't one in the past. Sequestering carbon by planting trees is a complex and disputed topic, and it may often be preferable for land simply to return to whatever kind of wild cover it had before being taken for agriculture. The land is not just being 'rewilded' to soak up carbon dioxide, but to protect and eventually restore biodiversity.

We still don't know enough to make all the right choices – on questions like biodiversity we don't even know half of what we need to – but we have already left the choices very late so we have to set priorities. Hitherto the strategy for cutting emissions, so far as there is one, has been to 'advance on a broad front', combining gradual changes in personal behaviour with slow but steady progress from industrial and agricultural sources. Except that there has been no progress: in no 'non-Covid' year since global heating was recognised as a problem has the amount of greenhouse gases being dumped into the atmosphere declined.

Perhaps we have failed because we cannot move fast enough as a civilisation. Yet if there is a sort of natural speed-limit on how fast human societies can fundamentally change – I'm not claiming that there is, but the thought does occur – then the appropriate response is not despair. It is to research and invest in specific workarounds that might just keep disaster at bay while the fundamental changes we need take hold.

* * *

Precision fermentation is potentially as big a breakthrough as farming was 10,000 years ago, because it makes food production independent of the climate – indeed, even independent of the land. If we can learn to live on and like 'food from the air', most of the constraints and vulnerabilities that came with agriculture shrink or just vanish.

The quickest and least controversial way to move to high-volume production of 'solar' food is by marketing it as animal feed. There is no 'ick' factor involved: the animals won't care that their food is microbial in origin, and volume production would quickly bring the cost down to a level competitive even with soya. That would kick-start the process of liberating large amounts of agricultural land for rewilding. It might even persuade governments to end the massive agricultural subsidies they provide to the biggest and richest producers. Spend the money instead on compensating those who are being forced out of the farming business by technological change. As far as people are concerned, there is definitely an 'ick' factor inherent in 'food from bacteria' (if we are being honest about it), and pointing out that we already cheerfully consume cheese, yoghurt and fermented foods made with the help of bacteria will not end popular discomfort with the concept (which would of course be amplified by the propaganda of those in the food industry who stand to lose from a shift to 'farm-free' foods).

Still, we really do want to get human beings using this food too. The likeliest avenue is via the expanding meat substitute industry that caters to vegetarians and vegans, where many products would benefit from more convincing sources of protein and fat. Thence to the baking industry, where the 'microbial' origin of the food protein would be equally invisible, and finally out in the open in a thousand different forms as a generation of cooks takes advantage of the possibilities opened up by new sources of both old and new proteins, fats and carbohydrates. This epochal shift in the human diet is probably the only hope for preserving the world's remaining biodiversity, its best hope for reducing food-related emissions in good time – and even a sort of hedge against a mass die-back of the human race or outright extinction.

Food and Global Disasters

All the potential global disasters that could plausibly threaten the human race with such a fate – a massive asteroid strike like the K-T Event 66 million years ago, a huge and prolonged volcanic eruption on the scale of the Yellowstone super-volcano (last eruption 70,000 years ago), or a human-caused 'nuclear winter' (any time between next week and never) – operate by driving enough dust up into the stratosphere to block out the sun's light and kill most plant life. Famine then kills off the animal life as well.

Really big asteroid strikes are extremely rare, volcanic eruptions large enough to affect the global food supply may occur once in a thousand years, and we have no experience of a full-scale nuclear war between the great powers, but the probabilities are certainly higher than zero in all cases. To them we can add the permanent risk of showers of high-energy cosmic rays caused by giant solar flares from our own sun: no blast of solar radiation powerful enough to burn out our electronic infrastructure has intersected with our orbit at exactly the wrong time since we became so utterly dependent on electronics, but such 'Carrington Events', although infrequent on a century scale, are essentially random and can occur at any time.[59] An eight-billion-population world whose means of communications and transport have been trashed would quickly become a very hungry world.

For all these reasons, it would be an excellent idea to have a food production industry that is close to the consumers, does not depend on the availability of sunlight, and is well protected from radiation surges, to see us through a possible period of years with severely reduced agricultural production. Indeed, that's exactly the sort of fall-back position a global civilisation that wants to ensure its long-term survival would be building right now even if it weren't facing a climate crisis.

8

NEGATIVE EMISSIONS

'Negative emissions' – more formally known as Carbon Dioxide Removal (CDR) – officially became part of the part of the IPCC's strategy at the 2015 Congress of the Parties in Paris, when it was tacitly acknowledged that cutting emissions alone would not suffice to halt dangerous climate change. Nobody said 'if we can't stop the idiots from dumping more CO_2 into the air, maybe we can at least slow the warming by taking some of it back out,' but that was the implicit logic. What they weren't ready to admit to the public (or, in some cases, to themselves) was that a lot of this CDR would be required permanently because it would likely be impossible to end greenhouse gas emissions entirely over the short or medium term – or even ever.

For example, we cannot stop using cement as a building material until an equivalent but carbon-neutral substance is developed, tested and deployed at scale, which could take fifty, seventy-five or a hundred years. So to achieve 'net zero' by mid-century – the consensus target for avoiding runaway warming – we will have to be extracting and sequestering at least as much CO_2 each year by 2050 as the concrete-making process is producing. Overall, the IPCC targets for Carbon Dioxide Removal are 75 million tonnes annually by 2030, 10 billion tonnes by 2050, and 20 billion tonnes by 2100. (Current removals: less than 1 million tonnes annually.)

The new emphasis on CDR in 2015 made good sense, and researchers began searching for the most effective way to take a lot of the excess CO_2 of human origin out of the atmosphere. Unfortunately, they backed the wrong horse.

BECCS

> To rely on this technique to deliver us from climate change
> is to demonstrate a degree of faith that is out of keeping with
> scientific rigour.
>
> **Tim Kruger, Oliver Geden and Steve Rayner, 'Abandon**
> **Hype in Climate Models', *Guardian*, 26 April 2016**

> If it sounds too good to be true, it probably is.
>
> **Folk wisdom**

'Bio-Energy with Carbon Capture and Storage' – BECCS – was all the rage in the years after the IPCC Special Report on 1.5°C Global Warming of 2018. It offered energy and large-scale CO_2 removal in the same package: first you grow plants that capture carbon dioxide, then you burn them to produce electricity – but the CO_2 that is released when they burn is captured and injected into permanent underground storage in suitable rocks. And that can theoretically be done over and over again on the same land, pulling a potentially unlimited amount of CO_2 out of the air while generating huge amounts of power. The IPCC report estimated that by 2100 BECCS could be removing a net amount of 3.3 billion tonnes of carbon dioxide from the air each year.

The big bet on BECCS at the Paris summit of 2015 and the Special Report of 2018 was a product of desperation. It was untried and under-researched, but promised to sequester carbon on a massive scale and to be largely self-financing (by producing electricity). However, BECCS did not stand up well under subsequent investigation.

The energy derived from burning a fast-growing plant like

switchgrass is only half of what you would get from an equivalent amount of coal; it might not even cover the costs of cultivating, irrigating, harvesting and transporting the plant material. More seriously, the amount of land required to make a dent in carbon emissions by using BECCS was ridiculous. It was calculated that growing dedicated crops for BECCS would require between a tenth and a quarter of a hectare of land per hypothetical tonne of carbon removed annually from the atmosphere. Therefore, to grow enough biomass and sequester enough carbon to have a 50 per cent chance of staying below +2.0°C would require taking over a land area between one and two times as large as India. That's between a quarter and a half of the world's arable land, and therefore roughly a quarter to a half of total current food production. The amounts of water and nitrogen fertiliser needed were similarly preposterous.

So we no longer hear all that much about BECCS. There will be specific purposes for which BECCS may still make sense. It's just not going to be a big part of the solution (if there is one).

Carbon Capture and Sequestration (CCS)

This was once the only idea for carbon dioxide removal. It simply proposed to strip the CO_2 out of the exhaust gases from burning fossil fuels and get rid of it somehow so that it did not enter the atmosphere. But it quickly became clear that the only way of sequestrating the captured CO_2 permanently was to pump it underground under high pressure into rock formations that would hold it securely for centuries or longer – and the cost of burying it turned out to be as high or higher than the cost of capturing it.

It costs $15–$25 per tonne to capture CO_2 from industrial processes producing 'pure' or highly concentrated CO_2 streams (like ethanol production or natural gas processing), but it can as high as \$40–\$120 for 'dilute' gas streams like those in cement, steel and chemical production.[60] Add in \$30–\$90 a tonne for sequestration, and the price was just too high for any commercial enterprise operating in a free market where competitors were not obliged to do CCS.

As a result, we had several decades of performative 'carbon capture', especially in the coal industry, which would announce a new pilot carbon-capture plant somewhere every five years or so to allay public concern about the industry's emissions. None of those pilots ever went mainstream or even reported on costs, but they would be discreetly removed from the stage after a while and a new pilot plant would open with much fanfare somewhere else. Which may explain why, after all these years, only 0.1 per cent of anthropogenic CO_2 is captured and sequestered by CCS.[61]

Ocean Iron Fertilisation

> We are not sure whether there will be any negative impacts of this process, but our calculations indicate that every year, if we cover about 2–3 per cent of the deep ocean surface with iron in this way, we will remove about 35 billion tonnes of greenhouse gases. I'd suggest with that one technology, it is possible that we could deliver a massive potential solution.[62]
> Sir David King, Chair, Centre for Climate Repair, University of Cambridge

Since total human emissions of CO_2 in 2022 were 36.7 billion tonnes, that would pretty well balance humanity's current account for carbon dioxide in one go. Problem solved! Well done! Next?

David King was rebooting an idea for cooling the planet that has been kicking around since 1988, when American oceanographer John Martin told a conference at at the Woods Hole Oceanographic Institution: 'Give me a half tanker of iron and I will give you an Ice Age.' This led directly to one of the earliest strategies for negative emissions: 'fertilising' the oceans with finely ground iron-rich dust.

Ocean Iron Fertilisation (OIF) provides iron, the key missing element whose absence restricts the growth of phytoplankton (microscopic algae that grow by photosynthesis in the sunlit upper levels of the ocean). The powdered iron enables the very rapid growth – a thousandfold or even a millionfold faster than without it – of immense 'blooms' of phytoplankton which absorb carbon dioxide from the water as they grow. They die off quickly (a few weeks), and the larger plankton sink to the bottom if they are not eaten, carrying the CO_2 they absorbed with them. The volumes involved are so huge that even a modest increase would be potentially relevant to the climate if that CO_2 stayed on the bottom.

It seemed like a good idea at a time when there were not many good ideas around. Payment for the sequestered carbon (once they figured out how to measure it) would allegedly have come in the form of carbon credits (never clear from whom) for taking all that carbon out of circulation semi-permanently. By 2008 venture capitalists were backing entrepreneurs whose websites talked grandly about 'forests in the ocean', and much powdered iron was thrown off the sterns of ships. Local phytoplankton blooms duly occurred, but the studies that would provide real data were never done. The start-up companies involved, having attracted considerable venture capital, folded their tents and vanished, and OIF remained on the shelf until the Centre for Climate Repair (CCR) at Cambridge University came along fifteen years later and resurrected the idea.

There had not been much useful research in the meantime,

as anti-ocean-dumping legislation originally introduced to curb the activities of the wildcatters also hampered genuinely scientific work. Key questions like how much CO_2 the phytoplankton will absorb as they grow and how much of that carbon will subsequently be carried down to the bottom and sequestered there when the plankton die remain to be determined.

Some of those questions might have been answered by a joint research project in 2022, when Cambridge University's Centre for Climate Repair and India's Institute of Maritime Studies got together to spread iron-coated rice husks over part of the Arabian Sea, but the experiment was demolished by a storm. New-style OIF could still be a game-changer if two conditions are met: that there are no negative impacts from putting a very large amount of iron (plus various nitrates, silicates and phosphates) into the sea; and that most of the carbon dioxide absorbed by all that organic matter ends up on the bottom and stays there. Those were the key questions last time too, and they still haven't been answered. OIF is still worth another look, but most of the theoretical and modelling work suggests that the return on investment would be poor.

Reforestation/Afforestation

Everybody's favourite kind of negative emissions is just growing trees, and the 2021 Glasgow summit included a side agreement between countries containing 85 per cent of the world's forests to protect the forest 'sinks' that absorb carbon dioxide. The 'Leaders' Declaration on Forest and Land Use' promised to end deforestation worldwide by 2030, and large sums of money were mentioned ($19.2 billion in public and private funding).

Preserving existing forests delivers immediate benefits in the form of avoided greenhouse gas emissions, and naturally attracts the bulk of the funding. Large-scale reforestation and afforestation projects (planting trees where there were once forests and on former grasslands respectively) only has long-term, one-time-only benefits, and will receive considerably less money. Explicit attempts to 'restore the land carbon sinks' – that is, full-scale rewilding of extensive areas as discussed in the previous chapter – is not yet on the international political agenda. Moreover, executing the 'Leaders' agreement is largely down to the individual countries that signed it, and there is no permanent secretariat keeping track of what is done and what left undone. It will do some good, but not nearly enough.

Direct Air Capture (DAC)

This is a technology in which giant fans draw air into contact with solids or liquids that absorb CO_2. The material is then heated to release the CO_2, which is eventually sequestered underground (in most cases). It is a batch process and requires a lot of energy, but, after about fifteen years of development, small pilot projects are giving way to the first commercial-scale DAC plants. Canadian-based Carbon Engineering, in partnership with 1Point5 and Occidental Petroleum, is building a plant capable of extracting a million tonnes of carbon dioxide a year from the atmosphere in the oil-rich Permian Basin in Texas, and is planning a DAC plant of similar size in north-eastern Scotland in partnership with UK firm Storegga and Japan's Mitsui (it'll be ready in 2026). In August 2023, the US Department of Energy announced a grant of $1.2 billion for two further DAC 'Hubs' of the same scale in south-western Louisiana (Swiss-based Climeworks and California's Heirloom will provide the DAC technology, Ohio-based Battelle the management), and South Texas (Climate Engineering and Occidental Petroleum again). There is at least $2 billion more in the department's budget for DAC, and it is funding feasibility studies for nineteen more Hub sites across the country.

These are the first million-tonne scale Carbon Dioxide Removal projects anywhere, and much will be learned. The cost per tonne of CO_2 sequestered will be \$250–\$600 in the early stages of operation, but the DoE launched a 'Carbon Negative Shot' initiative in late 2021 that aims to bring it down to \$100 a tonne within ten years. That would still leave DAC heavily dependent on government funding even for running costs; only governments can spend that kind of money on non-profit projects that serve the public good. And note that in this initiative, the Biden administration has deliberately created a precedent of global relevance, especially for countries with free-market economies: the message is that it is legitimate and indeed normal for governments to take the lead in funding and directing massive civil engineering projects associated with climate change.

It's looking increasingly likely that this technology will become the Direct Air Capture standard. The US National Academy of Sciences and the IPCC estimate that by 2050 the world's countries will have to remove 10 billion tonnes of CO_2 a year to stay at net-zero, which implies that 10,000 DAC hubs of the current million-tonnes-a-year scale would be needed. That's about \$2.5 trillion to build them (over more than twenty-five years) and a trillion a year to run them, so about the equivalent of current annual world military spending. And all you'd get for your money is a planet suitable for human habitation.

> Direct-air-capture technology gives our industry a license to continue to operate for the sixty, seventy, eighty years that I think it's going to be very much needed.[63]
>
> **Vicki Hollub, CEO, Occidental Petroleum**

Vicki Hollub's remark at an oil and gas industry conference in

March 2023 alarmed many environmental activists. Were the Carbon Dioxide Removal advocates making a pact with the devil? She had put Occidental's money down to create the partnership with Carbon Engineering (CE) that made the first big CDR plant possible, but some of the carbon dioxide that 'Oxi' puts into the ground will be used for 'enhanced oil recovery' (scavenging more oil from old fields), which is sacrilege to most campaigners. It didn't help when ExxonMobil CEO Darren Woods declared that DAC technology is 'the Holy Grail'. Such talk is increasingly common in the industry, which has some knaves in high places but few fools.

The writing is on the wall, and most senior people in the oil industry can read. Hollub's DAC project was a ground-breaking first in the industry, and it's hardly surprising that she needed to include some return on capital in the deal to sweeten the pill. It's noteworthy, however, that the Occidental-CE partnership's follow-on deal with the US Department of Energy to build and operate the South Texas hub involves no 'enhanced oil recovery': just pump the CO_2 into the ground, keep it there, *and get paid for it.*

This is the real deal: getting paid for it. Oil production will be plummeting by the 2030s, as fossil-fueled vehicles are removed from the road by legislation already on the books. However, very similar technologies will be required in the same areas to build and operate the 10,000 giant hubs around the world that must inject many billions of tonnes of captured CO_2 into secure rock formations underground each year for the foreseeable future. There's at least a trillion dollars a year in that, and 'first movers' get priority access.

Q: Why did the Vikings never lose a battle?
A: Because they always changed sides in time.

The oil and gas industry runs Louisiana. It's everything outside of New Orleans: the main industry, all those pipelines and refineries. It's fascinating because it's the belly of the beast. And yet the climate impacts in Louisiana are huge and happening now, and my department gets the most funding of any department at Louisiana State University because there's so many environmental impacts happening in Louisiana. And I have colleagues who talk about blue carbon and engineering these coastal systems to do more for carbon storage. These people who are chemical engineers, fracking experts – they care about Louisiana, about the world. They have the know-how, and they want to use their skills to put carbon in the ground.

I think this is the place it's going to be worked out, and these are the people who have the power. Like I live down the block from these big Exxon executives, right there in my neighbourhood. This is their state and they see what's happening. They don't want their state to go away and so they're invested in this. We just have to figure out how to connect the dots.

Cheryl Harrison, Assistant professor, Department of
Oceanography & Coastal Sciences, LSU

So are you determined to see your old enemies in chains, or would you rather win the war?

'When you sup with the devil use a long spoon,' warned the old adage, but it did imply that sometimes you do need to have dealings with him. For several generations oil industry executives have been cunning and ruthless enemies to the environmental movement and in particular to the climate science community, so the Greens and the scientists are profoundly suspicious of anybody who starts making deals with them. Caution is certainly necessary in such dealings, but you make peace with your enemies, not with your friends – once they are desperate enough to make and keep deals. Some in the oil industry are ready to do that in order to ensure their futures, and some haven't understood the nature of their plight yet, so the task is to know which are which. (No, I wouldn't trust Darren Woods either. And least of all, Sultan Ahmed Al Jaber, president of COP28 and CEO of Abu Dhabi National Oil Company (ADNOC)).

Enhanced Weathering

To start with, 'enhanced weathering' was about people grinding up mountains and spreading CO_2 absorbent rock dust on the land, but the quickest and cheapest way to draw carbon dioxide down from the air and bury it in rock ('carbon mineralisation') is to go to mines where the rock has already been ground up to get the valuable minerals out. Professor Greg Dipple of the Carbon Mineralisation Lab at the University of British Columbia has been working with mine tailings for two decades.

> I'm a geologist, and about twenty years ago, people started talking about sequestering carbon dioxide in mined rock. We started looking at mine wastes in various places in Canada and ultimately overseas, and found that the mined wastes were already turning into carbonate rocks on their own through chemical weathering, albeit at a very modest scale. This opened up the idea that we could figure out what was limiting the rate of the reaction, and once we understood the limits we could design approaches and technologies to make these reactions happen fast enough to have a meaningful impact on the climate problem.
>
> Greg Dipple

Not every mineral can absorb CO_2 quickly, but Dipple's team has found ones that do, and they have already done studies in an operating nickel mine in Western Australia and an operating diamond mine in Canada. In the latest tests they are getting up to the tonne scale of CO_2 captured, and no major energy input is required: just wet the right kind of rock (already ground into a fine sand by the miners) and stand back. Mines with the right kind of rock are often in out-of-the-way places, but taking carbon dioxide out of circulation has the same net effect on the atmosphere wherever you do it – and some of these mines are producing tens of millions of tonnes of finely ground rock a year.

> With really reactive material, which is not that common, we've calculated that we could be running this process at $30–$50 a tonne [of CO_2 sequestered], which is very inexpensive, but we're not going to get a gigatonne a year at that scale. Only a few million tonnes annually at best. But there are other technologies we can use to make the material more reactive. Those are going to be more in the $100 a tonne range, roughly competitive with rival approaches. So it's part of the solution, definitely into the hundreds of millions of tonnes. A gigatonne would be a hard push, but it's possible, given the scale of activities that are predicted to be occurring by 2040 or 2050.
>
> Greg Dipple

A gigatonne (the equivalent of 10,000 fully loaded aircraft carriers) is only around 2 per cent of our current annual emissions of CO_2 (40 billion tonnes), so this is no magic bullet. But it's one of the few possible CDR techniques that could be up and running in five years, and significantly large within ten. Every little bit helps.

Artificial Ocean Alkalinisation (AOA)

> We have big knowledge gaps. We have some hints that maybe ocean alkalinisation works well in model world. It looks like you can take up a lot of CO_2, but then we don't know what the biological implications are. So we're missing some critical information.
>
> David Keller, GEOMAR Helmholtz Centre
> for Ocean Research, Kiel

Many people are tempted by AOA, because it simultaneously addresses our worst problems in the air (CO_2) and in the sea (acidification). Granite and other silicate rocks naturally absorb carbon dioxide and remove it from the system, but they do it on millennial time-scales. We're in a bit of a hurry, and AOA speeds that process up by grinding the rocks very fine. If you take a cubic metre of rock and grind it up into tens of thousands of minute grains, you have vastly increased its surface area. But there are problems.

> We need a huge amount of rock. We would have to grind down twice the Matterhorn every year. Grind entire mountains down to powder, transport it to the ocean, put it in, and make sure it really reacts with the CO_2. For every ton of coal or oil or gas we burn, we would have to dissolve about four tonnes of rock [to neutralise the fuel burned]. The coal mining industry or the [hard-rock] mining industry would be happy to deliver that, and we already move that quantity of rock around today, globally, so it's not an impossible task, but of course it has huge environmental impacts.
>
> It will be dusty and dirty, and we have to see what happens if we put such an amount into the ocean. It could remove

too much CO_2 and make the oceans really alkaline. The zoo-
plankton could start eating dust grains instead of grazing on
plants, and that would not be good for the animals. It might
change the light level, so the water might become murky
and that's not good for photosynthesis. We have studied the
chemistry; that looks benign. But there are many side effects
we have to study very carefully before we can go ahead and
do it.

Andreas Oschlies, Head of the Biogeochemical Modelling
Research Unit, GEOMAR Helmholtz Centre
for Ocean Research, Kiel

In principle, it's almost elegant. The molecules of CO_2 in
the water are disassembled into carbonate rock and bicarbo-
nate ions that just float around: the ions are not a gas and they
can't escape back into the air. And the CO_2 prised apart by the
rock is immediately replaced by more CO_2 molecules pushed
into the water by the workings of partial pressure. The entire
ocean becomes a perpetual motion engine drawing CO_2 down
from the air and sequestering it harmlessly forever – and in the
process reversing the acidification that has also been caused by
the excess CO_2. Just keep adding ground rock.

Could it work as well in practice? That's very hard to believe.
Would it work at all, or would the side effects be prohibitively
bad? Nobody knows, but there are many assumptions in the
plan, each of which would be a show-stopper if it is wrong.
Moreover, how would all this be paid for? The amount of
energy the mining, grinding and transportation of rock would
probably require is equal to the total annual energy consumption of France or Korea, and since there would be no product
or profit, it would all have to be paid for by governments, presumably out of taxes. Yet AOA will remain on the agenda until

an equally comprehensive rival comes along, because people are getting desperate for a Carbon Dioxide Removal alternative that can deliver at scale.

Direct Ocean Capture

Recently there has been a shift in the thinking from grinding up mountains to less heroic methods of extracting carbon dioxide from seawater, most of which involve using electricity to separate the two. Furthest ahead is Captura Corp., a company spun off by the California Institute of Technology (Caltech) which has since 2022 been running a pilot facility on the shore in Newport Beach, California. Last year it began work on a far larger ocean installation. Essentially, the process mimics the technology of Direct Air Capture, but takes advantage of the fact that the CO_2 content of water is 150 times denser than that of air. Indeed, Captura calls its process Direct Ocean Capture.[64]

In the Caltech process, a small amount of seawater is diverted from the main intake stream for purification. This purified salt water is then split by 'electrodialysis' into an acid and an alkaline base. The acidic water is added to the primary flow of ocean water through the plant, triggering a chemical process that draws the CO_2 out (with the assistance of a gas-liquid contactor and vacuum pump). The carbon dioxide comes off as a pure stream and is stored, while the alkalinised water is returned to the acidic, now de-carbonised seawater in the system, neutralising it. The water is then returned to the ocean, ready to absorb more CO_2 from the air. The electricity is renewable, the carbon is captured, and the technology would scale up easily.

This technology can electrochemically enhance and restore the oceans' capacity for carbon dioxide removal from the atmosphere at a globally relevant scale, thereby mitigating ongoing and accelerating climate change.[65]

Gaurav Sant, Director, Institute of
Carbon Management, UCLA

One indication that this technology is on the right track is that at least two other teams are developing similar 'electrochemical' projects. The SeaChange group at the Institute of Carbon Management of the University of California, Los Angeles (UCLA) is working with a process that locks the CO_2 into solid limestone and produces green hydrogen as a by-product; they are opening pilot plants in Los Angeles and Singapore. A group at the Massachusetts Institute of Technology (MIT) led by chemical engineer T. Alan Hatton and mechanical engineer Kripa Varanasi is working on an 'electrochemical' process for extracting CO_2 from salt water that resembles the Caltech process except that it omits the expensive and fragile bipolar membranes. All three variants promise cheap removal of carbon dioxide, reduction of ocean acidification and scalability to the gigatonne level, and all envisage modules that can be located alongside desalination plants, ocean platforms (oil, wind etc.) and even large ocean-going ships.

All these rival outfits are admirably ambitious – UCLA researchers estimate that 1,800 of their planned full-scale plants would sequester 10 billion tonnes of atmospheric carbon dioxide per year, which is about a quarter of total annual human greenhouse gas emissions – but they are really in the earliest stages of the development process, still debating questions of electrode erosion and the like. In ten or fifteen years, this ocean-based technology, rolled out at scale, will probably be among our most important tools for limiting climate change and avoiding disaster. Direct Ocean Capture may even supplant Direct Air Capture as the CDR technology of choice. But the hard truth is that it won't be doing much for us until the 2030s, so it cannot rescue us from our geoengineering dilemma.

* * *

This extraordinary variety of potential 'negative emissions' techniques is encouraging, but it's also telling us something. We have already taken on the job, although we are not admitting it yet. This is the sort of work that Lovelock's 'planetary maintenance engineers' would actually do. We may choose to see these tasks as just emergency measures, 'transitional' to something better, and imagine we'll be able to stop in a few years, but that's delusional. All the tasks listed above, if they turn out to be effective, will need to be done for decades or even centuries.

And for the purposes of the debate about when we may have to start geoengineering, none of these negative emissions technologies could cancel out as much as 2 per cent of our annual greenhouse gas emissions by 2030. Later, when they have been deployed at scale, they can counter our current emissions and even begin working on our huge backlog of historical emissions.

But when it comes to our 2035 target, only one thing will really make a difference and that is how much we have cut our carbon dioxide emissions by that time. There's little chance that it will be enough.

9

THE SEA AND THE ICE

Just as a reminder of how much we still don't know, a scientific paper published in March 2021 revealed a previously unknown source of carbon dioxide emissions caused by human activity: bottom trawlers. Trawlers tow very large nets behind them to scoop up fish, and some of those nets are weighted so that they scrape along the bottom and catch the fish that live down there. It has been known for a long time that this causes cumulative damage to the seabed, which in heavily trawled areas like the East China Sea looks like it has had a clean shave (leaving no hiding places for juvenile fish). What was not known until recently is that the heavy nets also disturb the bottom sediments and cause the release of large amounts of CO_2 – about one gigatonne a year, which is as much as is emitted by the entire commercial aviation sector.

Little of the marine CO_2 released from the sediments in this way ends up in the atmosphere; it remains in the water and contributes to ocean acidification. But the higher the amount of CO_2 already dissolved in the ocean, the less able the ocean is to absorb more CO_2 from the air, so it all joins up and does the same damage. This raises two questions. One, what other major sources of anthropogenic emissions have we not yet found (because people are making big physical changes in almost every part of the planet)? And two, now that we know about this one, what can we do about it?

We cannot know the answer to question one, but question two is simple: all that's required to eliminate this huge release of carbon dioxide is to ban bottom trawling in about 4 per cent of the world's oceans, most of it within national waters.

Easier said than done, however: if you try you'll find yourself in a political war with the fishing industry, and that's an interest group that rarely loses.[66]

* * *

The 'world-ocean' covers more than two-thirds of the planet, and it contains at least half the biomass of living things – up to 80 per cent according to some estimates – so when it starts to change because of human activities, especially greenhouse gas emissions, there are bound to be many impacts on marine life and also on us. The three questions that really matter are: will the changes in ocean temperature and acidity stay within limits that do not radically alter the oceans' ability to sustain abundant life or trigger big changes (up or down) in the global climate? Second, can fish stocks stay big and diverse enough to go on playing a major role in human nutrition? Finally, can we avoid a rise in global sea level big enough to drown major coastal cities and flood fertile coastal plains?

Since the start of the industrial revolution, the ocean has absorbed about a third of our annual emissions of carbon dioxide, which has brought its pH down from 8.2 to 8.1. 'Down' means more acidic and the pH scale is logarithmic, so this represents a 26 per cent increase in acidity. This drop of just 0.1 pH has already had a significant impact on marine life and especially on shellfish, which are having increasing difficulty in forming their shells. A predicted further drop in pH of 0.3–0.4 (equivalent to a 100–150 per cent increase in acidity) by the last decades of this century would have a much bigger impact but we do not know exactly how severe the damage to the marine biosphere would be.[67] We also do not know how proposed Artificial Ocean Alkalinisation and/or Direct Ocean Capture measures may affect the acidity of the oceans.

Current research suggests that the maximum rise in average

global temperature predicted for this century will not cause catastrophic losses in fish population: between 5 and 17 per cent losses by 2100, with an average 5 per cent decline for every 1°C of warming.[68] We should also be aware, however, that some recent 'marine heatwaves' in the sea surface temperature, notably in the North Atlantic and the North Pacific but in some other regions as well, have been classed by the US National Oceanic and Atmospheric Administration as Category 5 ('Beyond Extreme'). Big surprises are possible: we understand much less about the behaviour of the ocean than we do about the atmosphere.

* * *

> If you stop killing sea life and protect it, then it does come back.
> Callum Roberts, Professor of Marine Conservation,
> University of Exeter

Assuming the ocean doesn't spin out in some unforeseeable way, what is the future for fish? There are no reliable figures for the global fishing catch before the twentieth century, but it may have been as little as five million tonnes annually before the invention of trawlers. First came sail-driven 'Brixham trawlers' in the early 1800s, but by the 1870s there were steam trawlers that could pull much bigger nets and catch ten times as much fish. With the advent of 'factory-freezers' in the mid-twentieth century – big ships that could travel to distant waters, capture up to 400 tonnes of fish every time they released their nets (and discard 40 per cent of it as unwanted 'by-catch'), mechanically gut and fillet the rest, and quick-freeze them – the total catch had risen to 30 million tonnes a year by 1950. It peaked at 130 million tonnes in 1996.[69]

At that point practically every major fishery in the world was being depleted, and various measures were brought in to restrict the catches in areas that lay within the 200-nautical-mile exclusive economic zones (EEZs) of coastal countries. As a result, the global fish take has fallen gradually to around 95 million tonnes and is still slowly declining.

> The gradual recovery of ocean fish stocks since the 1990s has been one of the few positive trends in wildlife populations, and it has been achieved mainly by enforcing rules that leave more adult fish alive to breed. 'Marine protected areas' allow entire ecosystems to recover, and they include 8 per cent of the world's oceans already. There are ambitious plans to expand them, but they simply displace the fishing effort to other parts of the ocean unless fishing fleets are put out of business, which is usually very difficult politically. For most areas, 'managed' fisheries are a better option, and most of the seas around the developed countries are already enforcing management practices like catch limits and closed seasons.
>
> **Villy Christensen, Institute for the Oceans and Fisheries, University of British Columbia**

There is still serious overfishing in some parts of the tropical oceans, but even that can be curbed by regulation. The trends that probably cannot be reversed are the inexorable loss of the world's coral reefs, the migration of fish species from their traditional homes to cooler parts of the ocean, and the huge shift in the balance of predators and prey fish.

As a diver, I particularly regret the loss of the coral reefs, but they have survived other major ocean heating events and come back after hundreds, thousands or millions of years.* Traditional fishing communities are devastated when their fish

---

* Some Red Sea corals are already showing a greater tolerance for heat.

migrate towards higher latitudes to escape the heat, but the fish will be back again as and when the oceans cool, a hundred or a million years from now. Even the normal relationship between predator and prey species may be re-established eventually, although that may require the disappearance of the human race.

> There will be fish in the future ocean, but the composition of fish assemblages will be very different from current ones, with small prey fish dominating. Our results show that the trophic structure of marine ecosystems has changed at a global scale, in a manner consistent with fishing down marine food webs.
>
> Villy Christensen

* * *

> We are eating bait, and moving on to jellyfish and plankton.
> Daniel Pauly, Sea Around Us Project, Institute for the Oceans and Fisheries, University of British Columbia

The 'trophic structure' describes the feeding relationships between different organisms in the food chain (i.e. who eats whom), with top predators at the highest level and phytoplankton, the primary producers that eat nobody and grow by converting sunlight into biological energy, at the bottom.

On land, human beings have drastically simplified the diversity of the biosphere and virtually eliminated most of the rungs on the trophic ladder: it's directly from grass to cows to people. In the oceans, that has not happened, but the biomass of large predatory fish ('table fish') high on the food chain has shrunk by two-thirds in the past century, while the biomass of small prey fish has increased.[70]

This shift has been driven by human fishing, which has concentrated on those big predators, thinning their ranks down the decades, while the smaller prey fish (sardines, anchovies, capelin, etc.) have benefited from less predation as the predator population shrank. That is what Villy Christensen means by 'fishing down marine food webs': as the traditional food fish grow scarcer, the human fishing effort has been forced to exploit the lower trophic levels more. Jellyfish fingers next stop.

Or maybe not. There is a significant possibility that the pursuit of wild fish in the traditional fishery may be largely replaced by fish-farming. In 2018, half of the 'crude protein production' from all marine and freshwater sources was already farmed 'fish' (including crustaceans). Three-quarters of those farmed fish were tilapia, carp and the like, produced in fresh water on land; only one-eighth of the 6.9 million tonnes of farmed fish were at the higher trophic levels (large ocean predators raised in marine cages).[71]

Could the production of these latter fish in captivity be expanded enough to reduce the pressure on their wild counterparts and restore the old trophic balance? Perhaps, but for that the 'Fish In/Fish Out' ratio needs to improve even further. FIFO really means 'little fish in/big fish out': it is a measure of how many tonnes of small fish must be ground up for fish meal and fish oil to produce a given tonnage of big food fish like salmon, and thirty years ago the ratio was dreadful: up to three tonnes in for one tonne out. Now, however, fish meal and fish oil make up a dwindling proportion of fish feed, which is increasingly plant-based (and could be an early beneficiary of 'precision fermented' microbial food in future). Most land-based aquaculture species are now net producers of fish, while cage-raised salmon and trout are net neutral, one in/one out. Overall, global 'fed-aquaculture' (all fish that are fed in ponds or ocean cages) currently produces three to four times as much fish as it consumes.[72]

> A couple of years ago, we did a global estimate for future aquaculture, looking at how far could you go with this. Just like we have intensive farming, in many places farming can be combined with aquaculture. There are great possibilities for future seafood supply coming from aquaculture. Aquaculture is an industry that will expand; I have no doubt about that.
>
> Villy Christensen

Global consumption of wild and farmed fish combined provides only one-fifth of the total protein (77 million tonnes in 2018) that people eat in the form of meat, fish, milk and eggs. However, as domesticated land animal stocks decline – and decline they must, under any plausible scenario that does not end in disaster – farmed fish have great potential to fill the animal protein gap. Indeed, despite all the human threats to the integrity of the ocean ecosphere, from microplastics to seabed mining, the oceans are in better shape than the land. There's just this little problem: they are expanding.

Rising Sea Levels

> If you think about the future and global warming keeps on going and sea level rise keeps on going, you have to ask yourself: what do we do? One alternative would be: everybody migrates. Compared with that scenario, building a dam far out to sea actually has a small impact on people's lives.
>
> Joakim Kjellssson

Joakim Kjellsson is a professor of meteorology at the GEOMAR
Helmholst Centre for Ocean Research in Kiel on Germany's
Baltic coast, a lot of which would be underwater if the sea level
rose by a metre. He is working with an IPCC estimate that the
Global Mean Sea Level (GMSL) could rise a maximum of one
metre by the end of this century, but he is well aware that this
is a very optimistic 'no-surprises' scenario. The European Environment Agency, using model simulations that include both a
high emissions scenario (what we are living in now) and fast
disintegration of the polar ice sheets, estimates a possible GMSL
rise of up to 2.3m by 2100 and up to 5.4m by 2150.[73]

The IPCC's 'assessment reports' contain all the data from
which the more extreme, lower-probability predictions are
derived, but the Executive Summary (all that most journalists
will ever read) is written in a process in which governments
and scientists directly negotiate the final wording. This is where
the long tail of low-likelihood, high-damage probabilities gets
amputated and the lowest common denominator (usually the
'median estimate') is what makes it into the text. In almost every
government there is constant competition for budget share
between necessary long-term investments (such as climate) and
short-term demands on spending that will generate immediate
and visible results, including at the ballot box. So in the final arm-
wrestle between scientists and governments over the Executive
Summary, the governments always seek to downplay the scale
and urgency of the long-term climate threat. They usually win.

But what if European governments took the high-end sea
level predictions of the European Environment Agency seriously? Can Europeans afford to ignore them and leave their
coastal areas unprotected? The implications of that for the
countries of north-western Europe around the North Sea and
its eastward extension, the Baltic, are grim. Kjellsson reckons

that at five metres of sea-level rise between 20 and 30 million people in eastern England, on the Channel coasts of France and Belgium, in the Netherlands, northern Germany and Poland, in Denmark, southern Norway and Sweden, in the three Baltic states, and even on the Finnish and Russian coasts around St Petersburg, would be forced to abandon their homes to the sea.

As the ports are inundated, there would be successive retreats inland, involving a huge loss of assets and a major burden of investment in new, makeshift port facilities (that would also have to be abandoned at a later date). So Kjellsson and Sjoerd Groeskamp of the Royal Netherlands Institute for Sea Research, considered what might be done to avoid that disaster. Their answer was the 'Northern European Enclosure Dam'.[74]

Dam the North Sea

JOAKIM KJELLSSON: We thought, 'What if we pool resources together and build one massive enclosure dam to protect all of northern Europe – one dam that goes across the English Channel and one that goes from northern Scotland over to Norway?' That way, you close off all the North Sea and you protect up to fifteen European countries with one dam.

GD: How much material would you need to build a dam like that?

JK: It's more than 600 kilometres' worth of dam, but for the most part the North Sea is quite shallow, so you wouldn't have to go all that deep except when you get to the Norwegian Trench, where the depth is up to 300 metres. When you're building a dam, the width at the base is somewhere around twice the height, so you can quickly figure out that that's going to be a lot of material...

The amount of sand you would need is about the same amount of sand that is produced in the world annually. Of course, you would build it over decades, but yes, it's an absolutely enormous amount of material – and, of course, enormous cost with it: somewhere in the range of €250–550 billion.

But if you spread the cost over twenty years and share it between fifteen European countries, some of them quite large economies, you realise we're talking a small fraction of 1 per cent of their combined GDP annually, so it's a feasible number compared to other things.

The ambition is staggering. The authors of the proposal are talking about a 161km dam running from western Brittany in France to Cornwall in England, and two more dams going 145km from northern Scotland to the Shetland Islands and then 331km from the Shetlands to Norway. This would cut all the region's main ports off from the sea, from Rotterdam and London to Stockholm and St Petersburg, unless locks deep and broad enough to let even the biggest ships through were built into the dams. The only alternative would be new ports built outside the dams, facing the open sea, and railways along the dams connecting them to the shore.

Then there's the question of what to do with the all the fresh water that empties into the 'enclosure' from big rivers like the Neva, Vistula, Oder, Elbe, Rhine, Seine and the Thames. They deliver enough fresh water to the North and Baltic Seas and the English Channel to raise the sea level by 0.9 metres a year in the enclosed area, so the whole enterprise would all have been in vain unless an equivalent amount of water were pumped out of the enclosure into the open Atlantic each year. Around 100 major pumping stations would be

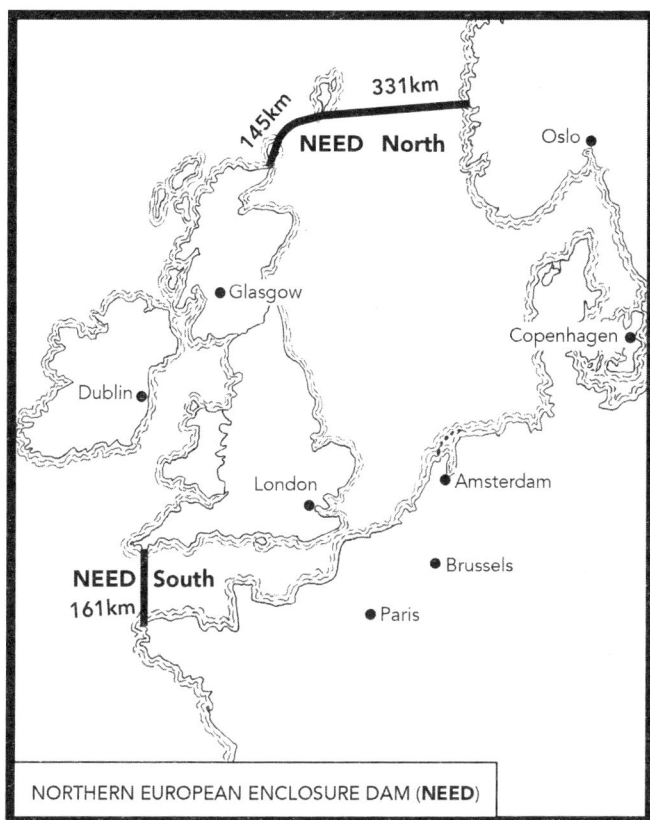

145km

331km

NEED North

Oslo •

• Glasgow

Copenhagen •

Dublin •

London •

• Amsterdam

NEED **South**

• Brussels

161km

• Paris

NORTHERN EUROPEAN ENCLOSURE DAM (**NEED**)

needed to transfer that volume to the ocean – and since the water being pumped out, far from the river mouths, would be salty, over a century the salinity of the water within the enclosure would fall by a factor of ten, devastating the existing salt-adapted ecosystems, biodiversity and fisheries. And the ocean currents that now pass through the English Channel and the North Sea would be forced to reorganise in ways that are hard to predict.

So is this a good idea? Of course not. Unfortunately, it may be the least bad remaining option if we let the sea level rise by even two metres. Similar schemes might be undertaken to enclose the Red Sea, the Persian Gulf and the Sea of Japan, as well as the Hudson River/Long Island Sound area, Delaware Bay, Chesapeake Bay and the Puget Sound/Salish Sea area, but neither Kjellsson nor Groeskamp want any of this to happen. They are just laying out the alternatives.

> The idea behind this is to do something that we call design fiction. We are proposing this to look at what the scenarios will be in the future if we don't mitigate the climate change threat... I pray that we never have to do it. I hope that people read about this and realise it's a crazy thing; that reducing our CO_2 emissions can be quite disruptive, but it's far less disruptive than all the other things we would have to do if we don't reduce our emissions.
>
> Joakim Kjellsson

Hang in there. Hope may be on the way – at least on the sea-level front.

Melting Ice

Changing sea levels are no big deal in planetary terms. Sea level has dropped by 130 metres and risen back up to about the present level more than twenty times since the onset of the current Ice Age 2.6 million years ago. In the more distant past, on occasions when all the ice on the planet melted, the sea level was up to 170 metres higher than now. But the coastlines determined by the current sea level matter a great deal to us, because that's where the human race has built most of its big cities and at least half of its major infrastructure.

So if you don't like the dam idea, how do we save our coasts from sea level rise? Well, stop the warming, obviously, but we know that's not going to happen fast enough to keep the coastlines where they are.

Assuming we are on course to melt all the ice on the planet, two-thirds of the sea level rise will come from the 2 per cent of the world's water that remains locked up in the Greenland ice cap and in the thick ice sheet that covers almost all of Antarctica in a normal inter-glacial period. The other third will come from 'thermal expansion': water expands in volume as it warms. It doesn't expand a lot: even a 5°C rise in average global temperature would increase the oceans' volume by only a quarter of 1 per cent (0.25 per cent). However, given the scale of the oceans, that's still an extra 3.5 million cubic kilometres of water, and it would drown many low-lying areas.

There's nothing we can do to prevent thermal expansion: as the world gets hotter, the oceans will expand. But maybe we *can* do something about the other two-thirds of the sea level rise that is due to land-based glaciers melting or just sliding into the sea (the melting of ice that is already floating on the water does not affect the sea level).

Of immediate concern in the next thirty to fifty years is the loss of, say, a quarter of the Greenland ice cap (two metres of sea level rise) and/or the collapse of a couple of big glaciers in West Antarctica (another two or three metres). Combined with the contribution from thermal expansion, we could be looking at a total sea level rise of as much as three or four metres as early as 2075. But there may be something we could do about that.

Stemming the Flows

Ninety per cent of ice flowing to the sea from the Antarctic ice sheet, and about half of that lost from Greenland, travels in narrow, fast ice streams. These streams measure tens of kilometres or less across. Fast glaciers slide on a film of water or wet sediment. Stemming the largest flows would allow the

ice sheets to thicken, slowing or even reversing their contribution to sea-level rise.

Professor John Moore, University of Lapland

'Stemming the largest flows'? The biggest glaciers coming off
the Greenland and Antarctic ice caps are more than a kilometre
high and many kilometres wide, they contain gigatonnes of ice,
and they move fast. The Jakobshavn glacier that empties into
Ilulissat Fjord on Greenland's west coast has increased its speed
to 12km per year since the floating ice shelf that held it back
disintegrated at the turn of the millennium, and it's still accelerating. How could anybody hold that back? Let's take it one
piece at a time.

Antarctica is an entire continent and Greenland is the world's
biggest island. Together they have coasts extending for almost
62,000 kilometres. Along much of those coastlines the ice sheets
reach right to the water's edge, but they are not losing mass at the
same speed everywhere. Like rivers draining a high plateau, most
of the 'run-off' from the ice caps takes the course of least resistance and follows specific 'stream-beds' down to the sea. Most of
the mass in the ice sheets moves only five or ten metres a year, but
these well-defined ice streams flow through the slower mass of ice
hundreds of times faster (1–5 kilometres per year and up).

The key is the way ice behaves as a material: the faster it flows,
the softer it gets. That tends to channel huge drainage basins
into relatively narrow, fast-flowing ice streams which are ultimately slowed down or stopped by high points on the seabed
grinding on the bottom of the ice shelf. The problem is that
the oceans are warming, and warm, dense, very salty water
is coming up and melting the underside of the ice shelves.
When they thin so much that they're no longer contacting

that seabed, there isn't the frictional force holding back the interior ice.

At Thwaites glacier in West Antarctica – the most vulnerable ice shelf on the planet, the major inflow of warm water is through a channel four kilometres across. If you could do something about that four kilometres, you could reverse the loss of that ice shelf.

John Moore

Thwaites, about the size of Florida, is sometimes called the 'Doomsday glacier', and the Amundsen Bay region it empties into is known as the 'soft underbelly of Antarctica'. If the Thwaites ice shelf came unstuck, the ice stream that feeds it could triple its speed. Some experts believe that this, in turn, could lead to the loss of the whole West Antarctic Ice Sheet.

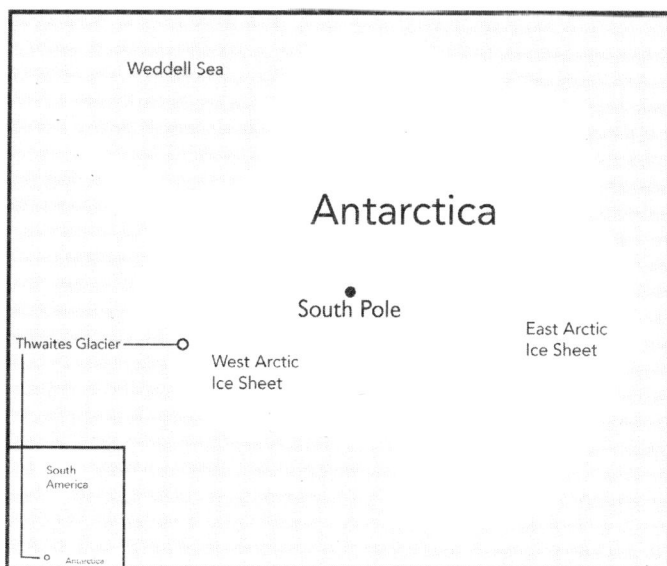

Conversely, however, if you could find a way to stop warm water from entering that four-kilometre channel and similar channels delivering warm, salty water to fifteen or twenty other big glaciers in Antarctica and Greenland, you could re-ground the glaciers and stop the rapid loss of ice – and the places where you would have to apply your solution is not the whole 62,000km of their coastlines; it's only the 200km or less where the ice streams reach the sea. Stop the accelerated melt there, and humanity might win an extra century or two to cope with all the other challenges posed by global warming without having to wage a continuous struggle to protect itself from rapidly rising sea levels at the same time.[75]

The real problem, in other words, comes down to stopping a couple of dozen glaciers ranging from 1–80km in width. So John Moore and a handful of other climate scientists began to indulge in some deliberately blue-sky thinking about how to stop a glacier. In a paper published in 2020, they put out a call for new ideas, listing all the techniques they could think of, however impractical or expensive they seemed. For example:

1. Drain the bed of the ice stream. Fast-moving glaciers slide over their beds on a lubricating cushion of water, so you might drill down through the ice and pump the water up to the surface (where it would refreeze). If the bed is dry, the glacier is stuck in place, and can only flow very slowly by internal deformation of the ice.

2. Create underwater obstacles that the floating part of the glacier will ground on, thus holding it fast against the shore and blocking the glacier's movement.

3. Build berms that prevent warmer seawater from making contact with the grounding line of the glacier.

And so on: at least a dozen good, middling and bad ideas. The one that seemed least unreasonable was an underwater dam.

> We initially thought, as stupid scientists, that we would just block [the 4-kilometre warm water channel] with rubble or dirt – you know, make a dam.
>
> The engineers laughed at us, slapped us around the face, and said 'Don't be so stupid. You don't want to have bull-dozers or something on the sea floor. You want to install pre-fabricated barriers, go in with a design that can be removed in case of unexpected consequences or to do maintenance.' So the idea is to have a concrete base as an anchor with a buoyant curtain attached. The buoyancy wants to keep the curtain vertical, the currents want to push it over, and there will be an equilibrium where it resists that flow.
>
> John Moore

The unsung genius who came up with the solution was Bowie Keefer, a semi-retired Canadian engineer with a strong interest in environmental problems (who didn't actually slap John Moore around the face).

> I was inspired by the kelp bed that we have off a beach here on Galiano [one of British Columbia's Gulf Islands]. A kelp bed consists of streamers of vegetation attached to the seabed and stretched upward by a buoyant bulb at the top. Paddling a kayak through the kelp, it occurred to me that if you attached those streamers together, you would have a curtain. And if we built a buoyant artificial curtain one or two or even 300 metres high, attached to the seabed by firm anchors, we could block the warm water from coming in and protect the ice sheet from being eroded from underneath.

Marine life, fish and whales could still go over the top, but the warm water [at the deeper level] would be blocked.

A very important aspect is that these ice sheets are shedding icebergs, and some of these are very deep. [Nine tenths of the mass of an iceberg is underwater.] If they float out and encounter the curtain, which would be made of numerous overlapping panels, the iceberg would push against the curtain, the curtain would bend, the iceberg would slide over the top, and the curtain would pop up again.

Bowie Keefer

The 'seabed anchored curtains' would be made of a reinforced tensile fabric: finely woven buoyant sheets of non-permeable, high-density polyethylene plastic (commercially available at about eight times the cost of normal plastic). The barrier curtains – not a single wall but overlapping panels – would extend an average of 300 metres above the bottom (500 metres in the deepest sections, which would require extra tensile strapping), and would divert the deep warm currents from the channels that lead to the underside of the glacier. They would, however, leave the top 200 metres just below the surface clear to allow free passage of cooler fresh water and floating ice. The exact height of the barrier would follow the interface between the warm, salty, dense current near the bottom and the cooler, fresher, less dense melt-water near the surface. An 80km run of these panels, to protect all of the West Antarctic ice sheet, would involve about 16 square kilometres of plastic, which would cost $10–15 billion to fabricate.

If we were trying this in a temperate ocean, we could do it with our technology tomorrow. Undersea cables, deep drilling platforms – large structures are moved around the seabed in water a lot deeper than we are proposing here.

Of course, special difficulties exist in Antarctica – extreme cold and storms, polar night, icebergs, no doubt about it, but we talked with engineers. The materials, the techniques for anchoring, the installation with large ocean-going ships all exist. They would need particular tweaks for working in Antarctica, but over a ten-year period you could install an 80-kilometre-long curtain that would protect Thwaites, Pine Island – the vulnerable glaciers in the Amundsen Bay region.

It would cost about $5 billion a year over ten years to install the curtain, and, after that, ongoing maintenance would be about $1 or $2 billion per year. That includes a fleet of five icebreakers, ten ice-reinforced ships, assembly in Punta Arenas, and the transportation.

John Moore

Fifty billion dollars over ten years is not a trivial sum, and of course there are other ice shelves that would need to be protected too, but it would be cheap at that price. The only alternative as the sea level rises will be to build far more expensive sea defences for hundreds of individual cities and towns, or a dozen mega-projects like Groeskamp and Kjellson's 'Northern European Enclosure Dam', or both. And even then, low-lying Pacific island countries would be drowned and low-lying coastal land all over the world – land that isn't worth the high cost of saving – would be lost. It would make more sense to deal with the problem at source.

There is some reason to worry that the stability of the Thwaites Glacier is already starting to fail: the 45-kilometre-wide floating ice shelf that slows the glacier's slide into the sea has developed diagonal cracks that may shatter it at some point in the next ten years. If the shelf goes, the speed of the glacier itself will increase and could even triple – and it's unlikely that scientists and engineers would be confident enough about their assumptions to take on the task of building Moore's 'seabed curtain' so soon.[76]

If the Thwaites ice shelf goes in the next ten or fifteen years, we'll just have to wear hip waders. But the few scientists who are active in this field are moving as fast as they can: John Moore is trying to win the support of indigenous Greenlanders for a geoengineering experiment aimed at preserving the biggest ice stream in Greenland: the Jakobshavn glacier that empties into Ilulissat Fjord.

This is much smaller than the big Antarctic glaciers, so the experiment would be on a more manageable scale – and, because it is contained within Greenland, the consent of only one government would be needed, not the twelve governments that have signed the Antarctic Treaty. On the other hand, it's tricky politically, because it's the Greenlanders who decide, and their priorities may be different.

> It's very important that we ask them: 'If you had control over your environment, the ice-fjord system, what configuration would you prefer? What would most benefit you in terms of fishing, tourism, ways of life?' Because Greenlanders, to be frank, don't care about the ice sheet. They don't use the ice sheet; they use the sea ice. They quite like that the summers are a bit warmer. They kind of like what the pluses of global warming are.
>
> We have to explain to them what the future would be for the ice sheet without any interventions, which would be a much more rapid retreat in the Jakobshavn Glacier outlet. We don't really know what would happen to the fisheries,

but [if they do nothing] how would the Greenlanders be viewed as stewards of the cryosphere, the ice sheet, by the rest of the world as small island states in the Pacific are lost, as Bangladesh and countries like that lose their land to catastrophic flooding? We want to show them the advantages of starting to value the ice sheet in ways that they've never historically done, but it's entirely up to them.

They may well say 'No, we're quite happy with the way things are now.' They make $35 million a year from the halibut fishery, so they need to be over-compensated for that kind of loss, to go from enjoying the halibut to 'We need to think about how we might better preserve the ice sheet for future generations.' That requires different funding mechanisms, such as 'payment for ecosystem services'.

There are payments for preservation of old-growth forests, things like the Amazon rainforest that people recognise as a global good. The ice sheet, we argue, and indeed the permafrost, should also be seen as global goods that deserve a funding mechanism. One mechanism might be insurance. I've spoken to insurance companies, and they don't know within a factor of a hundred what insurance premium level to set for Florida for hurricanes, because sea level rise adds remarkably to those risk estimates. If you could add a surcharge of something like $10 per year to an insurance premium in Florida, that could totally compensate the Greenlanders for being stewards of the ice sheet instead. Those are potential ideas that we want to talk to them about.

John Moore

The proposal could not entirely stop sea level rise: thermal expansion would continue, and so would an inevitable but relatively modest amount of glacial melt in areas beyond the reach of big geoengineering projects. However, it gives reason to hope that the IPCC's forecast of a maximum of 1.1-metre sea level rise by 2100 might turn out to be adjacent to the truth, not just an amiable fiction that ignores the factors they cannot yet include in their models – provided that the proposed method of containing the ice streams works, and is executed in good time.

What if the Permafrost Melts?

It isn't just glaciers that melt when it gets warmer. There's also the permafrost – the frozen layer of ground containing huge amounts of dead organic matter that underlies most of the land around the Arctic Ocean and a good deal of the adjacent sea-bottom as well. I spoke to Jessica Whiteside, an American associate professor of geochemistry in the School of Ocean and Earth Science at the University of Southampton in England. Normally she studies very ancient climates, tens or hundreds of millions of years ago, but this time she's looking at what you might call the recent past.

> **GD**: What are you looking for in Alaska, in this lake bed, and what will it tell you?
> **JESSICA WHITESIDE**: We're going to core lakes that encompass a climate range from the tundra to the permafrost – a ten-degrees-of-latitude range across Alaska. These lakes record archives of Earth's past for the past 12,000 years [after the ice cap retreated]. This is a key interval because the most recent time when we had a truly warm world, warmer than the present, was the Holocene thermal maximum 11,500 years ago. Temperatures were two to three degrees warmer than the present, and that's where we are projected to be going into in the year 2100. So it's a good partial analogue for future climate scenarios.

We know that when there's higher CO_2, the higher latitudes will warm the quickest and the most, but why? In the high latitudes, there's cold soils that impede the bacterial degradation of organic matter, so you end up with lots of material, dead plants and insects and everything just stored at the bottom of these lakes and wetlands. It gets compressed and it's there in the frozen ground, the permafrost – until there's a thaw.

Then that material is released out into the atmosphere as methane – and because methane is twenty-eight times more powerful than CO_2 as a warming gas, it results in much quicker warming in the Arctic region. For the very first time we're looking for a way to quantify the rate of methane production for 12,000 years ago, when the Earth was two or three degrees Celsius warmer.

The Arctic is now warming three or four times faster than the rest of the planet, and that's already having major effects on the climate even in the temperate parts of the Northern Hemisphere. It's therefore reasonable to ask if something like what occurred 12,000 years ago will happen again.

We don't know for sure if it was methane released from the permafrost that caused the sudden upward lurch in the temperature. Jessica Whiteside is trying to confirm this – but even if she does, it doesn't mean that it will play out exactly the same way this time. The surge of 12,000 years ago took place on a planet emerging from a glaciated state with an average global temperature 7°C colder than now. Would we see a similar surge, starting from the current temperatures, if more of the permafrost starts thawing? Climate scientists have to work through these questions one at a time, even though the answers are urgently required.

We're going out on inflatable boats and dropping core grab-
bers over the side that will pull the sediment up. It's just like
when you're building a layer cake. The top is the youngest
and the bottom is older, and it's a series of depositions that
gets compacted. Anything that's been living in this lake is
eventually brought to the bottom, so it will be a gooey core
that we will take out and bring back. We will look at different proxies or fingerprints of the past to understand how
methane changed during that time interval.

One thing we're looking at is fossilised fat particles, so we
can say what types of bacteria were there... There's a specific
biomarker for the methanogenic bacteria, so that tells us
something. We're also looking at the stable isotopes in a water
flea, a microscopic creature [that] eats both methanogenic
bacteria and other particles. Based on its diet, we'll know
when methane was in the lake.

There may be twice as much methane locked up in ice within
the permafrost zone of the Arctic than in the entire atmosphere,
and even more is stored in 'methane hydrates' (sometimes
called 'methane clathrates') buried under the sea-bottom sediments. The methane hydrates are stable as long as they are cold
and under high pressure, and the methane in the permafrost is
also safe unless it thaws. As the circumpolar region warms, one
of the scientists' greatest fears is that our current climate could
be overwhelmed and pushed into a 'hothouse' state by sudden
huge releases of methane from those sources.

I talked to Euan Nisbet, one of the world's leading methane
experts. He was unhappy about the fact that there is already
more than twice as much methane in the atmosphere as there
was before the industrial revolution – but he was remarkably
calm about the permafrost and the hydrates.

It's not just the Arctic. It's in, for example, the Amazon Delta. In the tropical oceans the hydrates are a kilometre or more down [so unlikely to escape], but in the Arctic they're only a couple of hundred metres down. If that warms up, you get gas release, and the danger is that it will get into the air. That's the bad side, and some of these hydrate resources are very large, so I pointed out a long time ago that the hydrates needed a lot of watching.

But we've done a huge amount of work on this. We still get very regular samples from [the Norwegian Arctic island of] Spitsbergen, and we've been getting samples from Alert in northern Canada, and one of our staff had a project finding gas vents in a couple of hundred metres of seawater.

Now, the nice thing is that even with a pretty big vent, the gas bubbles either dissolved in seawater or the methanotrophic bacteria ate the methane. So typically even quite big plumes of gas in a couple of hundred metres of seawater will rise a hundred metres but then they're gone. They convert to carbon dioxide, so you get somewhat acid water [when the CO_2 dissolves], but they don't reach the surface. That's the good side.

That's at sea. With the permafrost on land, the same thing is happening. In a lot of that permafrost, the methane is a couple of hundred metres down. But there again, as the bubbles of gas are coming up through the soil, they are eaten.

Professor Euan Nisbet, University of London

I was not expecting such good news: yes, some methane is escaping, but not very much, and certainly nothing apocalyptic. Perhaps I looked too cheerful, because Nisbet quickly informed me that there have been occasions in the past when vast deposits of methane hydrates have blown out – with disastrous results.

EUAN NISBET: If these big hydrate layers destabilise, particularly in places like the Amazon fan – a huge deposit off the river's mouth – or off Spitsbergen, where there's an enormous fan with a lot of hydrate in it, you get enormous submarine landslides.

GD: Like the Storegga Slide on the edge of the Norwegian continental shelf?

EN: The Storegga Slide was a bit over 8,000 years ago, around 6,300 BC, and that probably had a very big involvement of hydrate in it. That produced a 25-metre tsunami in the Shetland Islands and northern Scotland, and then a huge tsunami down the North Sea. It wiped out Doggerland in the middle of the North Sea, which was a reasonably big territory. It may even have been what produced the first Brexit by widening the Channel River into the English Channel.

GD: So the problem is more about landslides and tsunamis than dumping huge amounts of methane into the atmosphere?

EN: I don't want to be too reassuring. At the moment we're not seeing that signal, but the Arctic is warming more than anywhere else on the planet. So that's my answer right now, we're not seeing that signal – but that could turn around quite quickly.

PART THREE

CLIMATE GEOPOLITICS

*... in which the primitive state of political organisation
in a recently civilised primate species is discussed and
deplored – but does not necessarily have terminal
consequences*

10

WORST-CASE SCENARIOS

Governments with resources will be forced to engage in long, nightmarish episodes of triage deciding what and who can be salvaged from engulfment by a disordered environment. The choices will need to be made primarily among the poorest, not just abroad but at home. We have already previewed the images, in the course of the organisational and spiritual unravelling that was Hurricane Katrina. At progressively more extreme levels, the decisions will be increasingly harsh: morally agonising to those who must make and execute them – but in the end, morally deadening.

Leon Fuerth, former US diplomat and
National Security Adviser[77]

Leon Fuerth wrote those words in 2007, a time when the people who understood the climate problem were in despair. The first wind farms were appearing, but there were no other visible signs of hope. Today the public is far better informed, solar and wind power are competitive in price with fossil fuels, and new technologies good, bad and indifferent for cutting emissions or capturing carbon are spilling out onto the table begging to be considered. So why would anybody want to look at worst-case scenarios?

Because we are not out of the woods yet. So far, we have been dealing with the science of climate change. This is challenging enough because you can't negotiate with the laws of nature, but at least they aren't actively out to hurt us. Which is not something that can always be said about our fellow human beings.

What kinds of conflicts may arise as the warming affects global water and food supplies? Recent studies have tended to downplay the risk of large-scale war or even 'world war' *as a direct result of climate change*, but other very bad outcomes remain possible, maybe even likely.

The first-order impacts of the warming obviously include more evaporation from the oceans and more moisture in the atmosphere, which cause more destructive storms including more powerful hurricanes, typhoons and cyclones. Higher winds and torrential downpours mean increased flooding and more frequent landslides, with consequent destruction of lives, property and crops. Forest fires raging unchecked over large areas are also becoming a major problem.

The region where extreme heat alone is likely to threaten the lives of large numbers of healthy adults is the broad strip of territory that runs east from the Red Sea and the Persian Gulf across Pakistan and India to Bangladesh and beyond. It's well north of the equator, but it combines very high summer temperatures with very high humidity and very dense populations in a more lethal cocktail than any other part of the planet. This makes it extremely vulnerable to a rise in average global temperature, which will translate into a considerably higher rise in the local temperature over land in the tropics and the sub-tropics. The killer is the 'wet-bulb temperature', which is really a calculation of the combination of temperature and humidity at which it becomes impossible for human beings to cool their bodies by evaporation (perspiration). Even heat-adapted people cannot carry out normal outdoor activities past a wet-bulb temperature of 32°C, and beyond 35°C they start to die.

We project that the wet-bulb temperature will regularly exceed 35°C [in the Persian Gulf and northern South Asia] with less than 2.5°C of warming – a level that may be reached in the next several decades. Millions of people will be exposed to temperature extremes at the edge of and outside the range of natural variability in which our physiology evolved.

Colin Raymond et al., 'The emergence of heat and humidity too severe for human tolerance'[78]

Add three degrees to the average daily highs in cities from Dubai to Dhaka by 2040, assume that the humidity is still 80–90 per cent, and there will be a number of days in each of the summer months when the 'wet-bulb temperature' hits 35°C. That means mass casualties.

Many rich 'Gulfies' already decamp for the summer with their families to cooler venues like Los Angeles or London, while others with more modest resources go to Cairo or Istanbul. When it gets bad enough, many of these people will move permanently, having procured the relevant residence permits or citizenship, but hundreds of millions of others from coastal Arabia to Bangladesh will be exposed to the recurrent risk of killer heatwaves. Many of them may try to move then too, but they will find the gates of other countries barred against them – and for every person who needs to move because of the heat, there may be three to five more who need to move out of sheer hunger. (For the purposes of this discussion I am presuming that farm-free food technologies have spread very slowly or not at all.)

Population

It's around this point that the question of population usually comes up, and rightly so. Population is the key factor in determining the outcome of the climate emergency, for it is the consumption, waste products and other activities of human beings that are putting the pressure on the system.

WORLD POPULATION BEFORE 1804 IN MILLIONS

1 Billion

500 Million

100 Million

1804

500 BCE · 100 CE · 400 · 600 · 800 · 1000 · 1200 · 1400 · 1600 · 1800

WORLD POPULATION AFTER 1804 IN BILLIONS

8 Billion
7 Billion
6 Billion
5 Billion
4 Billion
3 Billion
2 Billion
1 Billion

1804 · 1825 · 1850 · 1900 · 1925 · 1950 · 1975 · 2000 · 2025

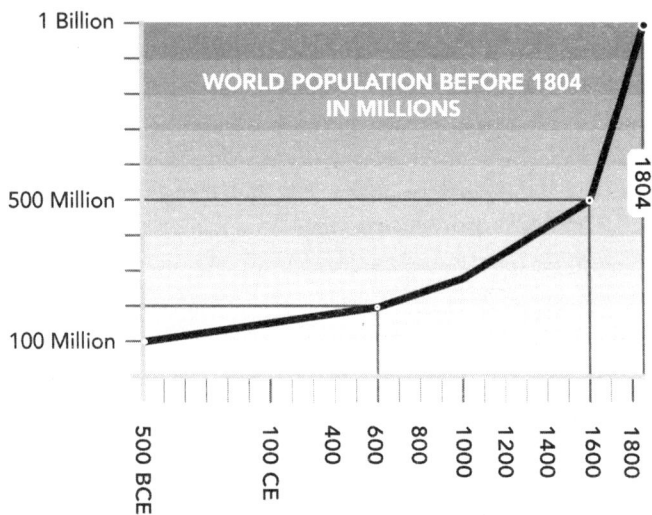

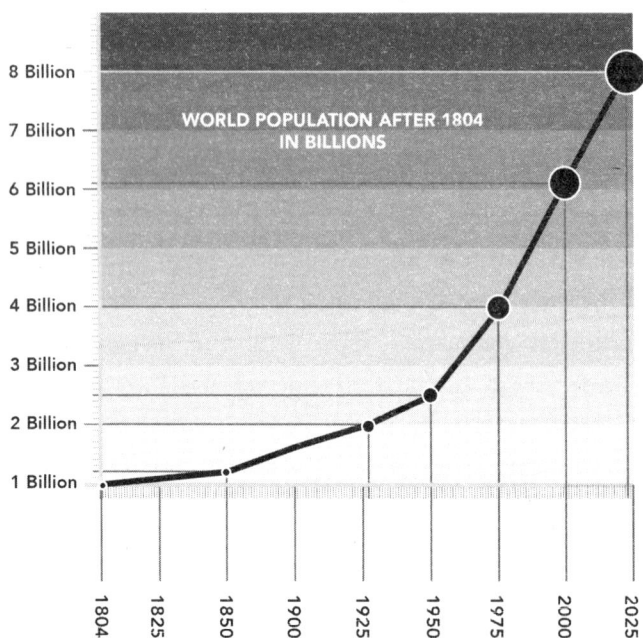

A global population of half a billion people, like that of the sixteenth century, could live just as lavishly and carelessly as we moderns do for many hundreds of years before their behaviour had a serious impact on the environment.

The ancestral human hunter-gatherer population, inhabiting only the Old World, probably never exceeded 10 *million*, because they were big-game predators who needed very large ranges. Once farming expanded the food supply, a slow-motion population explosion got underway: the human population had grown a hundredfold to one billion people by the time of the French Revolution ca.1800. But the next doubling, to 2 billion, took only about 130 years; the one after that, to 4 billion in 1975, took only forty-seven years. The doubling now complete, to 8 billion in 2023, has taken forty-eight years, so you could say that it's slowing down a bit. But it's still a lot of people.

If you increase the food supply for any wild species, its population will grow up to the new carrying capacity or beyond it. Human beings were essentially a wild population at the time of the agricultural revolution, and we were able to go on growing our numbers for a long time by continually taking over more land for crops and pasture. Then we got a second big boost to the food supply in the early twentieth century, mainly thanks to the invention of synthetic nitrogen fertiliser. (This is how we are feeding the second four billion.)

And then, just as the limits to that second, technology-fuelled expansion were looming on the horizon in the 1960s, along came cheap and effective birth control technologies, urbanisation, better education for women – and the birth rate started to drop. In ninety-one countries, containing well over half the world's population, births are already below the 'replacement level fertility' of 2.1 children per woman.[79] Indeed, in some of these countries the populations are already dropping.

China has almost one-and-a-half billion people now, but it will be down to 767 million by the end of the century. India is

on the same curve, but a generation behind China: 1.4 billion
now, peaking around 1.7 billion in 2050 – and back down to
one billion or less by 2100. This trend will not relieve the pressure on the environment in the short term, because at least one
more 'big generation' still has to finish growing up and forming
new households, with all the extra consumption that implies,
but around 2035–40 the pressure will begin to ease in most
parts of the world.

However, not a single country in Africa is below replacement level, nor any country in the Middle East except Lebanon
and a couple of the small, oil-rich principalities in the Gulf.
Egypt's population was 26 million in 1960, it's 102 million
now, and will be 160 million in 2050. That's a sixfold increase
in less than a century. Even more in Nigeria: 45 million in 1960,
206 million now, and 401 million at mid-century: a ninefold
increase. In fact three-quarters of world population growth in
the rest of this century will happen in Africa and the Middle
East, which together account for less than a quarter of the
human race today.[80]

Whether the world's population peaks at 9 billion, 10
billion or 11 billion later in this century depends almost entirely
on what happens to African and Middle Eastern birth rates.
They have now gone into a very gradual decline, but the latest
United Nations forecast still predicts that Africa's population
will almost quadruple to 3.44 billion people by the end of the
century. That will be more than Asia and around one-third of
the entire human race. But while this outcome would probably
entail a great deal of human suffering in Africa, it would not
have a critical impact on climate change outcomes. Nor would
high population growth in the Middle East.

Human population growth causes quite limited environmental damage if most people are either subsistence farmers or

poor city-dwellers. Producing food has a certain environmental cost that grows with the population, but high birth rates in Africa and the Middle East have meant that most people have stayed poor: even though the Gross Domestic Product may be expanding, it is not rising much faster than the population. And as long as most people in these regions remain poor, they will not be buying air conditioners, driving cars, eating large amounts of meat and generally consuming a great deal of energy, so their collective impact on global carbon dioxide emissions and other parameters of environmental concern will be far less than in other, richer parts of the world.

It sounds harsh to talk this way, but the fact is that a 25 per cent growth in the world's population over the next fifty years is not itself going to result in an immediate 25 per cent growth in the human pressure on the environment, and it is important to know this. Much more important from the climate perspective is what happens in the once-poor areas of Asia that have largely finished with population growth but are far from finished with growing their consumption of goods and services.

In the older developed countries, where there is slow or no population growth, emissions are very high but no longer growing: most people in these countries already have the typical First-World lifestyle, from cars and electronics to meat-rich meals and frequent holiday travel. There are roughly nine vehicles for every ten people in the United States, which is ridiculous, but it leaves little room for expansion. New-car sales in the United States have averaged only 15 million a year since 2005, mostly for replacement vehicles, with only modest annual fluctuations.

By contrast, new-car sales in China have gone from 5 million to 25 million in the same period (2005–19), and demand for electricity, domestic passenger flights and even meat consumption has seen comparable rises. In the rapidly developing countries, mostly in Asia, where disposable incomes are rising and birth rates are already near or below replacement level, the demand for the goods and services that have already reached saturation level in the rich countries is immense and so are the implications for the climate, whereas the consumption habits of large numbers of poor people do not cause major climate problems. China's share of global CO_2 emissions has grown elevenfold in the past fifty years, while the shares of America and Nigeria, at the two extremes of wealth, have remained largely unchanged.

Heat and Food

High temperatures will do more harm to crops in the tropics and the subtropics than in the temperate parts of the world. For the next generation, the most agonising blow dealt by global heating to human welfare will be the loss of food production in those regions – and the political and military consequences that may follow. To take just one example: the great river deltas (Nile, Ganges, Mekong, Yangtze, etc.) which together account for a disproportionate share of human food production, will be among the first areas to go underwater as sea levels rise. Here and elsewhere, those who bear the least historical blame for the warming will suffer the earliest and worst consequences. This will be a dominant and poisonous theme in the global politics of the next several generations.

There has been great progress in food production during the past sixty years, thanks mainly to the so-called 'Green Revolution'. The world's population has tripled since the Second World War and we're still managing to feed almost everybody, although the total amount of cultivated land has grown by no more than 10 per cent since 1945. But that bonus has now been paid out and there are debts falling due, especially in terms of groundwater supplies that have been pumped dry. We stand to lose almost as much food production in the tropics and the sub-tropics in the next two generations as we gained in the last two.

* * *

From now on we need to have a system where, for every 10,000 tonnes of carbon you emit, you have to take a Bangladeshi family to live with you.

Atiq Rahman, Executive Director, Bangladesh Centre for Advanced Studies, *Independent*, 20 June 2008

Bangladesh is only slightly bigger than England, but with three times as many people it has the highest population density of any large country (around 1,250 people per square kilometre). The country can just about feed itself from its own resources, for it is one of the most fertile places on the planet, but global warming will put an end to its self-sufficiency. In the parts of Bangladesh that won't be submerged by rising sea levels if we reach +3°C, it will be too hot to work outdoors for most of the day. The rivers will also be flooding as the Himalayan glaciers melt, and in any case the heat will be killing the crops. So Bangladeshis in their millions will try to move, as will people from other tropical and sub-tropical countries suffering shortages of food and unbearable heat. Their ideal destinations would be the countries of the temperate zones, further away from the equator, where food is still available and the heat is still bearable. However, Bangladeshis would be lucky even to get to India, which has already built a 3,400km-border wall, 3 metres high and guarded by 220,000 members of the Border Security Force, to stop illegal Bangladeshi migrants from crossing.

There would be a certain poetic justice if, as Atiq Rahman

suggests, the rich, developed countries far away had to take in refugees fleeing the effects of global heating in developing countries, because the rich countries have been burning fossil fuels on a massive scale for several centuries while the developing countries of the 'Global South' came late to the party. But there will be no climate justice: it is pure fantasy to imagine that the countries of the temperate regions would take in tens or hundreds of millions of climate refugees – or could feed and accommodate them all even if they were willing, because 70 per cent of the human race lives in the Global South. Indeed, the strategic analysts in all the traditional great powers are already considering how best to block such massive migrant 'threats'.

* * *

Can we put a number on the scale of the hunger? Such numbers are rare, mainly because governments don't want that sort of number becoming public, as it would certainly generate demands for actions that they are unable, unready or unwilling to take. This was the case with the confidential contracts given out by the World Bank more than a dozen years ago to non-governmental think tanks in the major world capitals, commissioning them to conduct studies on how much food production would fall in the country where they were based when the average global temperature reaches +2°C. The results of these studies have never been made public, presumably because the World Bank's member states forbade it. I only know of them because I interviewed the head of one of the think tanks that had the contracts.

> We looked at the impact of a two-degree C rise and the associated extra precipitation on the soil, on what is being grown today, how various districts would be affected, etc. For two

degrees of global warming, we found a 25 per cent loss of food production in India. One would expect large-scale hunger, large-scale migration and riots.

Dr Jyoti Parikh, Executive Director, Integrated Research and Development (IRAD), New Delhi

I asked Dr Parikh if she knew what the predictions had been for other countries. She only knew of one, for China, and only because somebody in the Beijing think tank, quite against the rules, had posted the executive summary of its report on China on that organisation's website for a few hours before the management (or the Great Firewall) took it down again. China would lose 38 per cent of its food production at +2°C, it said. No government that allows that to happen can hope to survive.

Peer-reviewed scientific papers do not predict such a big loss of food production at +2°C, especially for China, but research papers are inherently conservative whereas these studies may have been encouraged to examine plausible worst-case scenarios. However, even the Chinese figure is credible give the multiple hits the country would take in most warming scenarios: severe weakening of the north-east monsoon that brings the wheat-growing North Chinese Plain the greater part of its water; the exhaustion of the deep underground aquifers that provide most of the rest; the melting of the western glaciers that feed the great Chinese rivers; the more powerful storms and typhoons that will hammer the low-lying east coast; and the risk of inundation of the major rice-growing river deltas.

If the Chinese regime knows and believes that 38 per cent prediction, then no wonder it is investing more in non-fossil-fuel energy sources than any other country. True, at the same time China is raising its energy production by all available means, including even coal, to keep the economy growing, and that often leads to contradictory policy choices. The regime is no doubt well aware that a 38 per cent unemployment rate would be just as likely to cause its collapse as a 38 per cent fall in food production, so it must balance one risk against the other. And both China and India know that they can't just buy their way out of trouble.

> We have to remember that food will not come from outside. The whole world will be suffering food shortages. Today, if we have a food shortage, we can just go out and buy. This year we're importing, next year we'll be exporting; these kinds of things one takes in one's stride. But when the whole world is short, you cannot expect food to come from outside. It will be a very, very tough situation.
>
> Jyoti Parikh

Given the population disparity between the rich and the rest – two-thirds of humanity live in the tropics and subtropics – it's unlikely that any food surpluses still available in the temperate parts of the world would be large enough to meet the needs of a Global South experiencing simultaneous climate-related falls in food production in Asia, Africa, the Middle East and the Americas.

In an average year, both China and India (one third of the world's population between them) produce about 80 per cent of the food they need, and they also maintain between nine months' and a year's grain reserves, so they are far from the brink of famine. However, both they and every other government in the Global South knows that crop yields are likely to fall more or less in lock-step across the region as the temperatures rise – and that if the fall in production is more than temporary, food reserves will be used up fairly quickly. Then, if there's not enough grain available for sale internationally from temperate zone surpluses to cover the shortfall in domestic production in the tropics and subtropics, the highest bidder wins – and China can outbid everybody else.

Individual countries can sometimes ensure that everybody makes it through a time of food scarcity by imposing rationing, but there is no international authority to do the same: price is the rationing mechanism on the international grain market, and grain prices are subject to the same panicky fluctuations as oil prices. Does that mean waves of desperate refugees washing up against the borders of rich countries? That depends on which countries you are talking about.

* * *

There are already climate refugees on the move, mostly fleeing drought: young West African men whose family farms in the Sahel region have dried up and blown away, hoping to reach Europe and remit money back home to keep their families afloat (though these young men often drown in the Mediterranean instead); or 'caravans' of Honduran families who abandon their drought-stricken farms and try to make their way across Mexico and through the US border. There will be many more climate refugees as time passes, but for the foreseeable future this will probably be a problem restricted to places where a rich 'first world' country has a direct land border or an easily crossed sea frontier with a poor region where people are getting desperate.

The main points of friction will be along the US southern border and coast (Central American and Caribbean refugees), and in the southern tier of European Union countries (refugees from North Africa and the Greater Middle East from Egypt to Pakistan). Elsewhere, South Africa's northern borders, Australia's northern sea frontier and Russia's southern borders may also see some action, but the scale is likely to be much smaller because of the long distances involved. Many of the current generation of refugees are fleeing tyranny and/or war, but the number who are genuine climate refugees will rise inexorably. The number who find safe haven, however, will probably not rise proportionally.

Grand gestures like German Chancellor Angela Merkel's admission of almost a million migrants to Germany in 2016 will prove the exception. At some point, as pressure on the borders mounts, they will be closed – and contrary to popular belief, it's easy to close borders if you are really determined to do so. Just demonstrate your willingness to kill anybody who tries to cross illegally, and you probably won't even have to kill all that many.

To be clear: I am not advocating this approach as a way of dealing with mass refugee flows, but I am predicting it. There will eventually be so many people seeking refuge from the effects of the warming that letting them all in would stretch resources, and especially food, to breaking point, and such a policy will simply not command political support. The borders can be closed to refugees, and they will be closed. Indeed, this contingency is already being quietly discussed inside the governments of the richer countries, although never in public. In the United Kingdom, there's even an insider shorthand phrase for it: 'Lifeboat Britain'.

> What's the defining characteristic of a national lifeboat? First of all, it's got to stay afloat. It mustn't have too many people on it, or it will be in a position where 'we can't take you, we're going to sink'.
>
> James Lovelock

What Lovelock really meant by 'lifeboat' is a place that will keep its people alive when it all goes wrong, and the United

Kingdom could be such a place even in a world that has already passed +2°C. Maybe even past +3°C. The key criterion is whether or not you can still grow enough food where you live, and the British Isles probably meet that criterion. They are quite far north and surrounded by a cool ocean, so the rise in the local average temperature rise might be as little as half the rise in the global average. Producing enough food would certainly be a problem – the United Kingdom today imports just under half (45 per cent) of the food that it consumes – but it's a very fertile place so it could just about be managed if they imposed strict rationing and grew food on every square metre that isn't absolutely vertical. But there are already 68 million people in the country, and it's clear that the UK could not take in a lot more people in the midst of a crisis where food is already scarce – which brings us to the difficult question of lifeboat ethics.

GD: These national lifeboats are viable if the ratio of people to the amount of food you grow is still sustainable, but if you take too many on board, everybody starves. One can get into rather deep water quite quickly talking about this, but there is a limit and it would be enforced, wouldn't it?

JAMES LOVELOCK: You are absolutely right about lifeboat ethics and I've thought a lot about it – and my first thought is that this is the wrong time to ask. We're all so conditioned to try to save human beings – the most important thing of all – that we can't contemplate a world where we might have to select: 'Yes, you can come on board. No, you can't.' And I hope it doesn't come like that. I hope the planet does the selecting for us... It wouldn't just be the climate, there would be wars, there would be pestilence. The Four Horsemen would be riding, so it might not be that difficult for the 'lifeboat captains' to choose who could come aboard.

James Lovelock was a kind and civilised man trying to work out what would be moral and appropriate behaviour in an apocalypse. You'd hear a lot more people with a scientific background talking like this if they didn't find the subject so distasteful and disheartening. But it's always there, in the background, so let us pursue this topic to the bitter end. Where would these 'lifeboats' be?

JL: There was an event 55 million years ago when almost the same quantity of CO_2 as we're putting into the air now drove the global average temperature up 5°C or more. It may have reached 8°C higher in some regions, including the Arctic Ocean. They find the remains of crocodiles and animals of that type in fossils in the high Arctic from that time. Britain and Germany and in North America places like Newfoundland were tropical forests at that time. Further south, it was all desert.

There are other 'lifeboats' around the world. New Zealand is obviously one, and Tasmania, even tropical islands like the Hawaiian Islands may have tolerable climates. Always remember that 'tolerable' means that things will grow there. People can get by anywhere.

GD: Counting up all these places, Britain, Ireland, presumably Japan, Tasmania, maybe bits of Scandinavia... Does the Sahara reach the Baltic?

JL: It quite conceivably could, but the Arctic Ocean and Siberia and northern Canada and so on will certainly be quite productive in that [+5°] world.

GD: I'm counting up how many people you could support agriculturally in those regions, particularly given that the high Arctic isn't what you'd call ready to grow food plants. It's a long journey from permafrost to decent agricultural

land. We're talking maybe 200 million people could be supported that way?

JL: I'm afraid you could be right. I'd put it somewhere
between 100 million and a billion as the number of survivors
by the end of the century, if things go on as they're going.

These are contingencies and decision points that would only
become relevant in the latter part of the century, and even then
they will only become active if we fail to hold the warming down.
Let us delay further discussion of this until we have a better fix
on how likely we are to drift into such dangerous waters. In the
meantime, we can consider the geopolitical dangers that could
activated in the nearer term (say 2030–40) by warming that
has exceeded +1.5°C but has not yet reached +2°C – events that
would be large enough to disrupt the national efforts and international cooperation to avoid climate disaster that have already
begun and must expand rapidly and without interruption.
They come in three categories: refugees, wars and failed states.

Climate Refugees

The number of climate refugees can be expected to soar as the
climate changes intensify. The efforts of richer, less hard-hit
countries to stop refugees may eventually escalate from the
current 'push them back or let them drown' tactics employed on
the European Union's southern frontier in the Mediterranean
or the somewhat less brutal methods used on their borders by
the United States or Australia to actual shooting and killing,
but even if that happens these humanitarian calamities at
closed borders are very unlikely to lead to wars. The countries
that the refugees abandon would be far too weak militarily
to challenge the well-equipped armed forces of the intended
destination countries, and in any case there would be an

unspoken calculation in play in many governments in the source countries. If their former residents make it into the destination countries and find jobs, they will boost the flow of remittances back to the home country – and if they don't make it, at least they are no longer on the books.

There may not be wars, but there will be a real political cost: the great anger and hostility these tactics cause in the countries where the citizens fall victim to these tactics. Terrorism against the rich abroad and at home is bound to increase. More importantly, though, the ruthlessness of the rich countries in protecting their own borders at the cost of letting other people suffer and die will focus further attention on their historic responsibility for creating the whole warming problem that set those refugees in motion and will make cooperative efforts to cope with the climate crisis more difficult to organise and maintain.

Water Wars

The risk of climate-related 'water wars' in the developing world, especially between countries that share the same river, also receives a good deal of attention because the upstream country always has the option of hanging onto more than its fair share of the river's water when there's not enough to go around. Pakistan and India are the most frequently debated case, because the Indus river they share is vital to Pakistan and quite important to India, and it will be short of water once the Himalayan glaciers have melted. One can, at a stretch, imagine war over water between these two old enemies, but the fact that they both have nuclear weapons may paradoxically make them more cautious about full-scale war between them. (At least, that has been the general effect in similar cases elsewhere.)

Like Pakistan, Egypt is basically a desert with a big river

flowing through it, and it faces a similar problem with water security: the Nile's sources lie outside Egypt, partly in the Great Lakes countries of Africa but mostly (85 per cent of total volume) in the Blue Nile that flows out of the Ethiopian highlands. The construction of the Grand Ethiopian Renaissance Dam (GERD) on the Blue Nile has caused much anger and some anxiety in Egypt, but it is purely for power. No water will be extracted for irrigation, and once the GERD reservoir has filled there should be only a small loss of water going downstream to Egypt.

In every other case around the world, the upstream state enjoys such crushing military superiority over the downstream states that it can effectively do whatever it wants. Iraq and Syria might both be at war with Turkey today if they were still the moderately powerful and competent states that they were twenty years ago, for the Turks have built twenty-two big dams on the headwaters of the Tigris and Euphrates rivers, paying little heed to the fact that the Iraqis and Syrians downstream depend heavily on the water in those rivers to grow their food. The amount of water in both rivers reaching Syria and Iraq has fallen drastically (the downstream stretch of the Euphrates may dry up completely), and a severe, long-lasting drought has made people even more desperate. But Iraq was invaded by the US-led coalition in 2003 and Syria has been almost levelled by the civil war that began in 2011, so both countries are wrecks incapable of challenging Turkey in any meaningful way.

Much the same power relationship applies to the Brahmaputra River in Bangladesh and the Mekong River in Southeast Asia (Burma, Laos, Thailand, Cambodia and Vietnam). Both rivers have their headwaters on the Tibetan plateau in China, which is already building dams upstream and has plans for more. However, China is so much bigger and stronger than the downstream countries that it's hard to imagine any military conflict occurring no matter what Beijing does to the rivers upstream.

Pakistan vs India is the least incredible of the hypothetical 'water wars', but Kashmir is a considerably likelier *casus belli* for another Indo-Pakistan war than a dispute over the Indus driven by global warming. The risk of major wars in the Global South over climate issues is quite low; far more likely are numerous instances of state collapse.

Failed States

Keeping the population fed as the warming progresses will become a major challenge for many governments, and governments that cannot fulfil this basic duty tend not to survive. The contemporary emblem of this phenomenon is Somalia, where the original collapse of the regime was linked to a famine in the early 1990s which was followed by decades of bloody anarchy that is only partly under control even today. The dominant political phenomenon across the large parts of the Global South where the state is already weak – notably in the central belt of African countries – is likely not to be interstate wars but the proliferation of 'ungoverned spaces' where there is no effective political authority at all: no army, no courts, no schools, no healthcare, no police – just bands of armed men taking whatever is left. And awful though that would be, it would have no more effect on how the rest of the world fares than Somalia's plight has had on the world in the past three decades.

Hadley Cells

The poorer countries of the Global South will be the primary victims of climate change for the next generation or maybe even two generations, but that does not mean that the mid-latitudes – the temperate zones of the planet where most of the rich people live – will get off scot-free for the first half of this century. For some parts of the mid-latitudes, that may be true, but not for the parts affected by the expansion of the Hadley cells.

An atmospheric circulation system first identified by the early eighteenth-century English lawyer and amateur meteorologist George Hadley, the Hadley cells are responsible for tropical rainforests, the desert belts in both northern and southern hemispheres, and the trade winds that facilitated intercontinental voyages in the age of sail. The risk, in a warming world, is that the deserts will expand into the planet's major 'breadbaskets' – the US Midwest, the historic grain-growing regions of southern Europe and the Fertile Crescent, and the wheatlands of Australia and Argentina – cutting into the crop yields that currently provide most of the exportable grain surplus in the world. If the Hadley cells were to expand, much of that land would dry out and be lost to agriculture – and that is precisely what the Hadley cells are predicted to do in a warming climate.

The best estimate (and it's only an educated guess) is that the additional energy in the system with a couple of degrees of warming will shift the deserts' edge one or two degrees of latitude further away from the equator. That's only one or two hundred kilometres – but it's a 100 to 200-km-wide strip of the world's best agricultural land extending all the way around the world. Of course, it would never be as clear-cut as that in real life, but the direction of travel is clear, and so is the likely scale of the damage to food production. The developed world's agriculture will probably continue to be adequate to fill local needs, but it may not continue to produce very large surpluses for export. And there's another aspect of the behaviour of the atmosphere in the northern hemisphere, identified quite recently, that is also cause for growing concern.

> In the mid-latitudes, circum-global Rossby waves are associated with a strongly meandering jet stream and might cause simultaneous heatwaves and floods across the northern hemisphere... Rossby waves with wave-numbers 5 and 7... can induce simultaneous heat extremes in specific regions: Central North America, Eastern Europe and Eastern Asia for wave 5, and Western Central North America, Western Europe and Western Asia for wave 7... The regions affected by waves 5 and 7 account for a large fraction of global food production: for wheat, the United States, France and Russia produce 42 per cent, and for maize the United States and France alone produce 57 per cent.[81]
>
> **Kai Kornhuber, Earth Institute, Columbia University**

The northern 'jet stream'– a high-speed high-altitude atmospheric 'river' that flows from west to east all the way around the planet – marks the boundary between the polar air mass, sometimes called the 'Polar Vortex', and the warmer air masses of the mid-latitudes. (East-bound airliners crossing the North Atlantic and the North Pacific used to hitch a ride on the jet stream, saving up to an hour compared to the west-bound journey.) However, the energy that drives this circulation comes from the temperature difference between the two air masses it divides. As rapid warming in the Arctic shrinks that difference, the jet stream is slowing down.

Long S-shaped curves called Rossby waves were once just an occasional feature of the generally straight jet stream, but, as that has slowed both the frequency and the amplitude of the Rossby waves have increased. They are often now great loops deviating far north and south from the straight track, and those loops bring with them cold polar air rarely seen so far south, or warm air not often seen previously so far north. The waves are also showing a tendency to get 'stuck', with big loops containing very warm or very cold air loitering over the same area for several weeks – which can mean 'killer' heatwaves and extreme cold spells in unaccustomed parts of the northern hemisphere.

Extreme climate events like this have been tentatively linked to occasional changes in the jet stream – the 2003 European Heatwave, the floods in Pakistan and the heatwave in Russia in 2010, the Texan drought of 2011, the 'Great Freeze' of Texas a decade later – but there's still a lot of debate, with some claiming that the evidence for these perturbations of the jet stream does not rise above the level of natural variation. But I'll continue, with the proviso that the conclusions that follow are even more open to amendment than the usual scientific finding.

In June and July 2018, record-breaking heatwaves hit the western United States, Western Europe and the Caucasus-Caspian Sea region at the same time, while there was extreme rainfall and flash flooding on the US east coast, in eastern Europe and in Japan. Big Rossby loops that had taken over the entire jet stream (on the map it looked like a huge river meandering across a flat plain), and all the loops, containing hot dry air or cool moist air, were 'stuck' for upwards of two weeks. Norway had its hottest temperature ever but only half the normal rainfall for July, while in Japan severe floods and landslides caused by heavy rain destroyed more than 10,000 houses. Researchers started looking into it seriously, and within a year they had found that the same 'wave-7' pattern of stalled highs and lows over the same regions, extending all the way around the planet, had occurred during extreme weather events in the summers of 2003, 2006 and 2015.[82]

In all, the researchers found seven times in the past twenty years when this stalled wave-7 pattern lasted for more than two weeks – whereas in the previous twenty years (1980–2000) it had not happened once. This strongly suggests that these wave patterns may become more long-lasting or more frequent or both as the rapidly rising temperatures in the Arctic lessen the difference between the cold polar airmass and the warmer mid-latitude air masses, slowing the jet stream down. That would be very bad news for the world's food supply if it meant that summer weather in the 'breadbaskets' becomes more extreme, but it's early days yet and more research is needed. At the least, however, it serves as a reminder that from time to time 'unknown unknowns' are bound to emerge when the climate system is already so volatile – and that when they do, we may not immediately recognise them.

Artificial Winters

It's hard to imagine a plausible scenario in which the great powers go to war with each other over a climate issue. Opponents of geoengineering often suggest that putting sulphates or other reflective aerosols into the stratosphere could trigger such a war, but they underestimate the degree to which the nuclear-weapons powers are self-deterred by their knowledge of what a nuclear war would do to their own countries. Mistakes and accidents are always possible, and the great powers remain quite unreformed in their readiness to use 'conventional' military force against weaker countries (including sometimes against the proxies of rival great powers), but they are reluctant to risk direct military clashes among themselves. More than three-quarters of a century have passed since the first (and last) time that nuclear weapons were used in war.

The discovery during the 1980s, just before the end of the

Cold War, that an all-out nuclear war between the great powers would cause a 'nuclear winter' may have further strengthened deterrence. It still might not qualify as an 'extinction-level event', but it meant that those who survived the thousands of fireballs and the fallout, even those in countries that had not been directly targeted in the war, would spend the next decade of their lives freezing and starving in the semi-darkness.[83] But the 'proliferation' of nuclear weapons to lesser powers – Israel, India, Pakistan, North Korea – caused no similar alarm, as they had too few weapons to cause a nuclear winter. Or so everybody thought.

> If a conventional war starts [between India and Pakistan], anything could happen. We will fight, and when a nuclear-armed country fights to the end it will have consequences far beyond the borders. It will have consequences for the world.
>
> **Former Pakistan Prime Minister Imran Khan to the UN General Assembly, 27 September 2019**[84]

Whether the proximate cause of another Indo-Pakistani war was climate-linked or just another round of the struggle for Kashmir, the risk of escalation to nuclear weapons would be very high, for two reasons. Firstly, most of the nuclear-capable aircraft and missiles of the two sides are still vulnerable to being destroyed on the ground in a surprise attack. Secondly, the two countries are so close together that only a very brief warning time is available. In these circumstances, a policy of 'launch on warning', with all the risk of mistakes it entails, is the only rational option for the commanders on both sides, no matter what their publicly declared strategy may be.

The Indian and Pakistani nuclear arsenals each amount to about 150 warheads, growing by maybe a dozen a year on each side. That's a quite modest total compared to the superpowers, and many people still assume that the effects of any future nuclear war between the two – they have fought three full-scale conventional wars and half a dozen major skirmishes since independence – would be largely confined to South Asia. It was climate scientist Alan Robock's group at Rutgers University that determined in 2020 that this was not the case.

The first and worst victims would be Pakistani and Indian civilians, because cities will be on the target lists: that's where most of the major ports, airfields and critical infrastructure are. But those burning cities would loft enough 'black carbon' into the stratosphere to create a shroud of soot over the whole world within a few weeks. It wouldn't be the full-dress nuclear winter of superpower war, with 'darkness at noon', but it would dim the sun enough to drop temperatures and severely damage crop yields in the main food-producing regions of the planet.

The models made by the climate scientists did not concern themselves with 'prompt fatalities' in an Indo-Pakistani nuclear war (generally assumed to be in the low hundreds of millions for a 150-warhead strike on each country). They focused on the global climate effects of 300 nuclear explosions in the Indian subcontinent, most of them airbursts over cities. The main effect would be a severe drop in the average global temperature, and the consequence of that would be a comparable decline in global food production – with the worst-hit areas being in the northern hemisphere north of latitude 30°N (i.e. north of almost all of India and Pakistan). It may be counter-intuitive, but that's the way the climate system works.

The planet's most important 'breadbaskets' – grain-growing areas that produce a big crop surplus for export – are all just north of 30°N: the United States, Canada and Europe including Russia. Northern China's grain is mostly consumed at home, not exported, but it produces a comparable amount of food and it is in the same latitudes. The dimming effects of an Indo-Pakistani nuclear war would cause an average drop of 1.8°C over all the continents, with more cooling in the northern hemisphere and less in the southern, but in the key regions of North America, Europe and China it would be between −1.5°C and −4°C colder. The maximum cooling would be reached in the fourth year after the war and would continue for several more years before gradually returning to normal by around year fifteen. Post-conflict rainfall would follow much the same pattern of decline and recovery, falling by as much as half over the mid-western and north-eastern United States and for a time virtually ceasing over India and central China.

The collapse in food production would similarly worsen over three or four years, reaching 20–50 per cent below the pre-war norm in the 'breadbaskets' of the Northern Hemisphere, but the international grain trade would stop almost at once as the major grain-exporting countries of the northern hemisphere began stockpiling their surpluses in order to protect their own populations from mass starvation during the peak famine years. The southern hemisphere would suffer less, since there is a fairly limited interchange between the two hemispheres in the stratosphere, and the southern 'breadbaskets' – Australia and Argentina – might even still be able to export some grain, but they are small compared to those of the North and would not be able to make up for the huge shortfalls of food in the poorer parts of the northern hemisphere.[85]

There is no reliable way of estimating the post-conflict population loss in regions outside the Indian subcontinent during this lengthy ordeal, but it would probably exceed the immediate deaths from the actual nuclear war, and it would certainly distract everybody from the far greater threat of global heating. Scrabbling for food in the cold and the dark does not focus people's minds on their long-term problems. But when the effects of the local nuclear war in the Indian subcontinent finally faded, it would be right back to that bigger climate crisis in a significantly worsened form, because carbon dioxide would not have stopped accumulating during the hungry years.

None of this has yet happened, and as time passes the risk of a nuclear war in the subcontinent will diminish: experience with the operation of a mutual deterrence system will increase, and as more of the two countries' nuclear weapons are moved out to sea in submarines the pressure to 'launch on warning' will decline. Moreover, there are no potential calamities of the same gravity lying in wait in other parts of the world, even among countries that have nuclear weapons. North Korea's nuclear weapons are so few, for example, that a US nuclear response to a North Korean attack, should that improbable occasion arise, would not approach the level at which there would be significant climate effects.

Most of these hypotheticals are either low-probability, or distant enough in time that we can still prevent them. It could be worse.

11

THE GLOBAL SOUTH

A global problem requires global negotiations, and the fault line in the negotiations about climate divides the relatively 'developed' rich countries of North America, Europe, northern Asia (Russia, Korea and Japan) and Oceania (Australia and New Zealand) from the generally poorer 'developing' countries of what we now call the 'Global South'. This fault line coincides in most places with the line between former imperial powers and formerly colonised countries, but for climate purposes it divides those that have been burning fossil fuels on a large scale for a long time (*and are rich mostly because they did that*) from those that have only recently started down the same road.

We're not really talking about geography here. Indeed, most people in the 'Global South' live north of the equator, in places like China, northern India, Egypt and Mexico. The point is that they all have the same grievance: that the old, rich countries had used up most of the planet's capacity to absorb excess greenhouse gas emissions without toppling into runaway heating even before the countries of the Global South started down the road to industrialisation and prosperity. In recognition of their legitimate grievance, they were all given an exemption from the obligation to curb their own emissions in the very first international treaty aimed at curbing global heating, the United Nations Framework Convention on Climate Change of 1992 (UNFCCC). At the time it made sense, because the developed countries were responsible for 80 per cent of human greenhouse gases in the atmosphere (some had been burning fossil fuels on an industrial scale for two centuries), and they could and should clean up their own mess.

When the developing countries that have made some progress in building their own economies (and emissions) over the
past thirty years are now asked to accept an equal share in the
task of ending those GHG emissions and repairing the climate,
their response, understandably, is unenthusiastic. They retain
their exemptions even today – yet they must cooperate, or
everybody goes under together.

* * *

The countries of the Global South are equally blameless in terms
of historical responsibility for global warming, but their future
trajectories have dramatically diverged. China's emissions have
tripled in the past three decades and are now larger than the
total emissions of all the developed countries – or perhaps one
should say of all the other developed countries, for by most
standards China is now a fully industrialised country. Similarly,
India alone emits more GHGs than the entire European Union,
although nobody would claim that it is a typical developed
country of the Global North. Both retain their exempt status
– but they are even more vulnerable to the various impacts of
climate change than countries of the Global North. In mid-May
2022, the temperature in Delhi reached 49°C.

> India is likely to have 400 million people under such heat
> stress that 25 per cent of the time they would not be able to
> work in northern parts of the country. That's a very large pro-
> portion of man hours lost, or lives put at risk if they have to
> continue working in the heat. This is true today because a
> large proportion of India's labour – about 45–50 per cent –
> is still employed in farming, which is all performed outside.
> Rajat Gupta, Head of Sustainability, McKinsey Asia[86]

Most countries in the Global South will be taking big hits from wild weather earlier and more often than the countries of the Global North: the South is hotter, and therefore there's more energy in the system. However, we all risk busting through the climate's never-exceed limits on temperature if the big emitters of the Global South don't take control of their emissions. In one way or another, most international climate disputes are linked to this issue. The rest usually have to do with the even more controversial issue of 'loss and damage'.

The poorer countries have been demanding talks about 'loss and damage' for the past decade, as the damages and the losses that they have incurred from increasingly extreme weather pile up – and the countries of the Global North have systematically put off discussing the issue because they don't want to be held responsible for virtually unlimited damages. This is one of the reasons why for long periods global climate talks so often run aground. However, the commitment of the rich countries to transfer $100 billion a year to the countries of the South by 2023 to cope with climate change and the creation of a 'loss and damage' fund at the COP28 in Dubai in December 2023 are tacit acknowledgements that 'climate justice' cannot be ignored.

Energy Politics

There were two significant inflection points in the history of fossil fuels that have shaped the international energy politics of the twenty-first century. The first was after the end of the Second World War, when mass consumption societies rapidly emerged in the industrialised societies of the West and in a couple of East Asian countries. In 1945, only 40 per cent of American families owned cars; by 1980, almost 90 per cent did, and the first 'two-car families' were appearing. Annual world energy consumption from fossil fuels went from 20,000 terawatt-hours in 1950 (a terawatt is a trillion watts) to 70,000 terawatt-hours in 1980. Coal accounted for two-thirds of those terawatt-hours in 1950; by 1980 oil provided more energy than coal and gas together.[87]

The second inflection point came in the mid-1980s when the growth in energy use in older industrialised countries slowed ('market saturation'), while simultaneously it took off in Asia, home to more than half the world's people. We are still living in this second consumer boom, as the emission statistics clearly show. From 1985 to 2019, energy derived from fossil fuels almost doubled from 74,000 terawatt-hours to 136,000 terawatt-hours, with the great bulk of the expansion coming in Asia. And coal made a big comeback, since the biggest rapidly industrialising countries all have plentiful reserves within their own borders. (China, India and Indonesia now produce about two-thirds of the world's hard coal.)

From the dawn of the industrial revolution to 1990, humanity's total carbon dioxide emissions were 784 billion tonnes. Since 1990 (after the second inflection), we have emitted another 831 billion tonnes: more in the past three decades than we did in the previous three centuries – and all at a time when we already knew that those fossil fuels were causing dangerous global heating.[88] That's why it has been so hard to reach a global agreement on how to address climate change, and especially on how to make a fair allocation of the remaining rights to emit CO_2. Indeed, the fact that there has been no explicit attempt to negotiate emission rights for various countries within a limited global emissions budget, despite the self-evident need for such an approach, illustrates just how divided countries are.

The older developed countries have high but slowly declining emissions and account for about a third of total global emissions. However, China and India alone, still officially 'developing countries' and therefore exempt from controls, account for another third of all global emissions – and their share of emissions is rapidly growing. So, naturally, many people elsewhere blame China (and to a much lesser extent India) for wrecking the planet.

> They [China] still think it is their right to poison the planet because Europeans have done it in the past. This is a mad idea, as if they are saying: 'The West has brought us to the brink and now we have arrived to push everyone over the edge.'
>
> Mohamed Nasheed, former prime minister
> of the Maldives[89]

China may well feel resentful, because it was the only traditional great power that missed the industrialisation bus, and as a result has been playing catch-up (very successfully) for decades. Pretty much everybody else is fed up with China's slow progress on cutting carbon emissions: the country burns more than half of the world's coal, and even now it is building a new generation of coal-fired power stations.

This doesn't make for happy relations or easy negotiations, and some would even say (in private) that China is the biggest single obstacle to a faster global pace in cutting emissions. Others would say that's merely a convenient excuse for other countries to drag their own feet. Both accusations are true, but in a broader climate perspective the rest of the world should be grateful to China.

* * *

For at least two thousand years China, which accounts for nine tenths of East Asia's population, has been the only consistent peer rival to the 'West' (Europe and its daughter countries overseas) in wealth, power and technology. The industrial revolution began in Europe and China fell far behind for a couple of centuries, but it was bound to catch up with the West again eventually. Most of East Asia's early industry was destroyed in the wars of the 1930s and 1940s, so once peace was restored (1945 for Japan, 1949 for China, 1953 for Korea), a period of rapid growth was hardly surprising. How it happened, though, and particularly when it happened, was very surprising.

In 1950, per capita Gross Domestic Product in all three East Asian countries was under $100 a year. By 1985, it was $10,000 in Japan and $7,000 in South Korea. Other developed industrial countries had also gone through a high-growth phase of forty or fifty years when cheap labour was flooding into the cities to work in the new factories, but never before had there been a three-decade burst of 10 per cent per annum economic growth and energy use. However, the People's Republic of China somehow missed this bus too: in 1985, a long generation after the revolution, GDP per capita in the Chinese People's Republic was still only $300 a year.

There was no profound structural or cultural reason why the Chinese should lag behind the Japanese and Koreans, and indeed the breakaway Chinese province of Taiwan turned in a respectable performance despite a large burden of refugees and huge defence expenditures: by 1985, it had a GDP of $3,300 per capita. You can't even blame the failure on Communism: the old Soviet Union had a comparable period of high-speed growth between the end of the Civil War in 1920 and the German invasion in 1941, and the second generation of Chinese Communist leaders, from Deng Xiaoping to Hu Jintao, belatedly delivered the statutory East Asian three decades of 10 per cent growth for their country in 1985–2015. (Under Xi Jinping, economic growth has fallen back to a normal developed-country rate.)

If China's burst of growth had occurred at the same time as Japan's and South Korea's, then by 1985 one billion Chinese people would have been enjoying a GDP per capita of $10,000. We know what that means in terms of CO_2 emissions, because $10,000 is exactly China's GDP per capita right now, and it is currently responsible for more than 27 per cent of total world emissions.

In a different world, where the Chinese economy had grown at the same rate and *at the same time* as Japan's and Korea's, China would have accounted for a quarter of global emissions by 1985. Back then we had none of the new energy technologies to replace fossil fuels that we have now: no solar arrays and wind farms, no battery technology suitable for electric cars, no clue about possible ways to remove carbon dioxide from the atmosphere and sequester it safely and permanently. Climate science was in its infancy and could not have persuaded most people to understand the full measure of the danger we faced until it was far too late. We would probably have crossed the never-exceed $+2°C$ threshold by the late 1990s, perhaps still not grasping that there was such a threshold. And now, in the 2020s, we would already be trapped on the pathway leading to 'Hothouse Earth'.

Without noticing it, we have had a narrow escape, and we owe our good fortune to just one man. Mao Zedong single-handedly prevented the Chinese economy from taking off for a quarter-century with his ceaseless, furious political campaigns, from the 'Great Leap Forward' (1958–62) to the 'Cultural Revolution' (1966–76). Tens of millions died needlessly, and the country stayed very poor.

China is not poor any more, and its enormous and still growing emissions are the biggest single obstacle to the 'prompt and deep cuts in greenhouse-gas emissions' that everybody says they want. There is therefore a temptation elsewhere to see China as The Problem, but that is not fair: the Chinese are doing no more than the older rich countries have already done.

More to the point, they have already made a huge sacrifice by (involuntarily) postponing their industrialisation for more than thirty years. An entire generation of Chinese suffered grievously as a result, but perhaps we should put up a modest statue somewhere to Mao Zedong: The Man Who Saved the World.

While we're at it, maybe we should also light a candle for independent India's first prime minister, Jawaharlal Nehru. He dominated Indian politics from 1929, led the country to independence in 1947, and remained in power until his death in 1964. By that time, his democratic socialist values – mostly derived from the Fabian Society in Britain – were enshrined in the political and economic structure of the country. That meant state-owned and operated enterprises, especially in heavy industry, discouragement or strict regulation of private enterprise, high taxes, rationing and Five-Year Plans. Essentially it was Communism without the tyranny and the bloodshed, but definitely with the poor economic performance.

In terms of rapid industrialisation, both countries shot themselves in the foot. As late as 1990, India's per capita GDP was still higher than China's, but by 2000 China's was twice as high. Now it is five times higher. That's mostly because India was fifteen years later than China in breaking with the old ways, but it still aspires to become a fully industrialised country too, and it will achieve that goal even if it takes a little longer.

As for China, its true economic growth rate has probably already fallen to the developed world norm of 1–3 per cent, and demographic factors (a rapidly ageing and shrinking population) guarantee that it will stay down. China continues to burn more than half the world's coal and is adding more coal capacity – the lights must stay on – but it will also have more than half of the world's wind and solar power by 2025. They are trying hard.

India may never hit a 10 per cent growth rate, but it will continue to grow fairly fast until it achieves at least a Chinese level of prosperity, which means that its emissions are unlikely to peak until some time in the 2030s. It is building a lot of renewable power, but it does not envisage reaching net zero until 2070, twenty years after most other countries.

Typical hunter-gatherer short-termism. Western oil barons, Chinese Communist Party bureaucrats, Indian hyper-nationalists – they're all the same: first we'll make ourselves (or the country) rich and powerful; then we'll fix the damage. But look on the bright side. At least we get three staggered peaks in emissions, one for Europe and North America in the 1980s, one for China arriving in a few years' time, and one for India that is yet to come. That's a lot better than a single massive peak that would have arrived long before we were scientifically or technologically equipped to deal with it, and would have pushed the Earth System over the edge ten or fifteen years ago. We should be grateful for our good fortune and be patient with those who paid for it.

PART FOUR

DESPERATE SCIENTISTS AND LAST-DITCH IDEAS

... in which a means of bridging the gulf between aspiration and reality is proposed, its merits debated, and a recommendation made

12

A SHORT HISTORY OF GEOENGINEERING

Matt Watson, Professor of Volcanoes and Climate in the school of Earth Sciences at the University of Bristol, was one of the first scientists I went to see when I began this book, because he was involved in the concept of Solar Radiation Management even before it had acquired a name.

MATT WATSON: I don't think there is any way we'll stay at an increase of +1.5°C [average global temperature]. It's absolutely impossible; we'll be lucky to stay under +2.0°C. There's a lot of stuff built into our systems that will prevent us from keeping a lid on this and climate engineering research will become more important over time. There's always this separation between stuff that tries to draw carbon dioxide out of the atmosphere and the slightly blunter tool of controlling incoming radiation. Honestly, none of them are perfect. There is no silver bullet.

GD: But there are two things we know about Carbon Dioxide Removal (CDR). It's expensive, and it's slow-moving compared to Solar Radiation Management (SRM).

MW: Absolutely, and there are questions about scalability. I am, however, very glad that people are working on it, and probably it is at least part of the solution. But... if everybody on the planet was going to die when we reach +1.5°C, we would be thinking about SRM, not CDR. We would have no choice in the matter. There is certainly no way cdr can keep us under +1.5°C. It might be able to bring us back, but if we all dropped off a cliff at +1.5°C, it would be SRM or bust!

GD: Doesn't this lead you to the conclusion, not that we must do it now, but that SRM will have to be seriously considered for deployment?

MW: Absolutely. I hope it never happens, but there is definitely a point at which it would be immoral not to act, and that point is not as far off as I would like it to be... Some of the projections of +3.5°c, +4.0°c require us morally to research this even though I find it in places abhorrent, terrifying. People always pick me up on that word, but for the record I do find it absolutely terrifying.

But it may well end up being the least bad option. I don't think that's a particularly unlikely pathway, in all honesty... Nature is precious and special and as someone who revels in nature and considers themselves a natural scientist, it is difficult for me to imagine a world where we are controlling the climate. But my intuition is that in some number of decades, if we're not doing this, either we've had some incredible revolution in the way we behave as a species, or...

GD: Mass die-back.

MW: Yes, we're in real trouble.

Could we at least do something about the name? Solar Radiation Management? As Tim Lenton put it, it's 'as if a bunch of accountants were running some kind of climate control system.' Ken Caldeira pleads guilty.

I worked at Lawrence Livermore Labs up until 2005 and they had a programme they called 'carbon management', about sequestering CO_2 underground. Then I moved to Carnegie Science, and it partnered with folks at NASA's Ames Research Center to hold a meeting on geoengineering. Pete Warden put a document into headquarters and it hit some tripwire; people were worried about adverse media and public relations consequences if it came out that NASA was holding a meeting on geoengineering. We were in a meeting, so I said, 'Let's call it the most bureaucratic-sounding thing we can think of, something that sounds like jargon and will pass under the radar.' I remembered the 'carbon management' thing at Livermore Labs, and said: 'Why don't we call it *solar radiation management*?' We just put that on the title, sent it through to Washington again, and some bureaucrat rubber-stamped it without a question.

Ken Caldeira, Carnegie Institution for Science

The whole idea of 'Solar Radiation Management' came from volcanoes, which with every big eruption put sulphur dioxide into the stratosphere and cause some short-term global cooling.* By 2006, concerned scientists were realising that the response to their warnings about global heating was going to be much too little, much too late. It was already sixteen years since the Intergovernmental Panel on Climate Change's first assessment, but human greenhouse gas emissions had not declined nor even stalled. The world's energy consumption went up every year, and not even the share of fossil fuels in the global energy mix had declined: it was still more than 80 per cent of all power generated – and it still is today.

Most scientists assumed that people would eventually wake up to the danger when the climate damage became unbearable, but they knew enough to suspect that by then it would be too late to do anything useful about it. However, in scientific circles

---

* In recent years, with the emergence of other practical proposals for blocking some small part of the sun's radiation, Solar Radiation Management (SRM) has become a generic term used to embrace all of them. The specific technique of putting aerosols in the stratosphere by means of aircraft has been renamed Stratospheric Aerosol Injection (SAI).

research on climate engineering measures, even as a last-ditch
defence against runaway heating, was strongly discouraged,
and public discussion about it was utterly taboo. 'Not in front
of the children' was the policy, as if the public would instantly
lose any interest in reducing greenhouse gas emissions if it
heard about SRM.

Professor Mark Lawrence is an American-born scholar who
has pursued his scientific career in Germany since 1996.

> It was a fearful taboo: most researchers didn't want their
> name associated with the topic and so they wouldn't look
> at it. There were a handful of researchers who were really,
> really critical and sort of repressive about it, and made it
> very clear that they would more than look down on people
> who were working on the topic. There were some very
> nasty things said to Paul and about Paul in the scientific
> community.

'Paul' was Dr Paul Crutzen, a Dutch atmospheric chemist
working in Germany who had concluded that since no emissions cuts were happening anyway, there was nothing to lose by
opening up a discussion on the possible techniques and consequences of geoengineering the atmosphere to hold the heat
down. He was well aware that the only permanent solution to the
warming problem is a complete halt to anthropogenic (human-
caused) greenhouse gas emissions, but stopgap measures might
well be needed to contain the warming while human beings,
having woken to their peril, finally got their act together.

Crutzen was the right man to lead the charge, because he
was already a big gun in science. He shared the Nobel Prize
for Chemistry in 1995 for his work on the ozone hole, and
in 2000 he successfully proposed a new term, the Anthropo-

cene Epoch, to describe the period starting around 1950 when human beings began to substantially alter the Earth's surface, atmosphere, oceans and systems of nutrient cycling. So in 2006 he, Mark Lawrence, and four other scientists, including Ralph Cicerone, President of the US National Academy of Sciences, persuaded the journal *Climatic Change* to devote an entire section to the question of geoengineering. Crutzen led with his seminal article 'Albedo Enhancement by Stratospheric Sulfur Injections: A Contribution to Resolve a Policy Dilemma', and the five other scientists wrote supportive pieces advocating research on the topic.[90] It was possibly the most understated revolutionary manifesto in history: most people wouldn't even have known that 'albedo enhancement' was scientific language for making the planet reflect more sunlight back into space, i.e. Solar Radiation Management (SRM). But it didn't just open the subject up for debate; it blew the doors off.

> Geoengineering holds forth the promise of addressing global warming concerns for just a few billion dollars a year.
> **Newt Gingrich, Speaker, US House of Representatives**
> **(2008)**

Right-wing politicians in the United States used the new information (new to them, at least) in exactly the way that the scientists who tried to stifle any public discussion of climate engineering had predicted. People like Newt Gingrich now alternated between denying human-driven climate change outright and touting geoengineering as a magical Plan 'B' that would make the climate problem go away without requiring any changes in lifestyle or any cuts in fossil fuel use. But letting fear of people like Gingrich set the scientific agenda is self-defeating. Crutzen faced a fire-storm of criticism and abuse for going public about the potential of geoengineering, but he did the right thing.

> Crutzen and I wrote an article on the tenth anniversary of that special issue looking at breaking the taboo and asking the question: was it the right thing to do or not?... as a scientist I would generally rather make an informed decision than an uninformed decision, and we were heading with that taboo towards a lot of uninformed decisions.
>
> [Breaking the taboo] opened up a space where researchers could use good scientific methods instead of having hand-waving debates about what the implications of atmospheric interventions would be, and they managed to find out an enormous amount about what would work and what wouldn't work and what kinds of side effects could be expected.
>
> Mark Lawrence, scientific director, Institute for Advanced
> Sustainability Studies (IASS), Potsdam

In his 2006 article, Crutzen advocated research into how to hold a rising global average temperature down by reflecting a small portion (around 1–2 per cent) of incoming sunlight back into space before it reached the ground. This might be done by lofting a certain amount of sulphur dioxide into the stratosphere, since it was already known that large volcanic explosions did this, and that the gas they put up there had a temporary cooling effect without having any disastrous side effects on living things. And unlike all the other methods for reducing incoming sunlight that had from time to time been proposed (such as giant mirrors in space), it was imaginable that SRM with sulphur dioxide might accomplish this at a reasonable cost without needing any drastically new technology.

By far the preferred way to solve the policy-makers' dilemma [about how to stop the warming] is to lower the emissions of greenhouse gases. However, attempts in that direction have been grossly unsuccessful. While stabilisation of CO_2 would require a 60–80 per cent reduction in current anthropogenic CO_2 emissions, worldwide they actually increased by 2 per cent from 2001–02, a trend which will probably not change... Therefore, though by far not the best solution, the usefulness of artificially enhancing Earth's albedo might again be explored and debated as a way to... counteract growing CO_2 emissions.

Paul Crutzen, 'Albedo Enhancement by Stratospheric Sulfur Injections: A Contribution to Resolve a Policy Dilemma'

Everything Crutzen wrote in 2006 is still true and relevant today. Even the rate of increase in 'current anthropogenic CO_2 emissions' is about the same: the average rise in the three years 2017–19 was still around 2 per cent a year (so doubling every thirty-five years). Exploring the sulphur dioxide idea again would therefore be a good plan, especially since it is pre-tested, in the sense that nature already does it on a fairly large scale.

Volcanoes big enough to put large amounts of sulphur dioxide (SO_2) into the lower stratosphere (15–20km high) occur every few decades, but the first to benefit from close scientific observation was the eruption of Mount Pinatubo in the Philippines in 1991. The explosion blasted an estimated 17 million tonnes of sulphur dioxide into the stratosphere, where within weeks it combined with water vapour to form an aerosol of sulphuric acid particles that reflected enough incoming sunlight to lower the average world temperature by half a degree Celsius for an entire year. If human beings desperately needed to hold down or lower the global temperature for a while, this could be a useful method.

Crutzen was aware of a concern among scientists that loading the lower stratosphere with a volcano-scale amount of sulphur dioxide over the long term might cause unacceptable damage to the ozone layer. Sulphur dioxide in the stratosphere does not directly attack ozone, but sulphates can act as a catalyst and speed up the destruction of the ozone in the presence of the chlorine compounds (CFCs) that caused the ozone holes in the first place. However, Crutzen predicted that by the time anybody would want to resort to such an extreme measure, the amount of CFCs left in the upper atmosphere would have fallen to a harmless level. (CFCs were banned in 1987 by the Montreal Protocol, and the amount in the stratosphere was already falling by 2006.)

After Crutzen's paper, the taboo on theoretical research rapidly evaporated, and since then hundreds of scientific papers have been written on every aspect of the proposal. Yet after almost twenty years, despite the rising concern about global heating, there has still been no experimentation with small-scale releases of SO_2 in the stratosphere, mainly because a vocal minority continues to oppose even very small-scale tests in the atmosphere on the 'slippery slope' principle.

The SPICE Project

Not long after Crutzen's paper, there was a government-funded feasibility study in the UK on whether it would even be possible to put the necessary amount of sulphur dioxide up that high. This was the SPICE project – Stratospheric Particle Injection for Climate Engineering.

The first task of the SPICE scientists was to decide whether sulphur dioxide was really the best chemical for the job, given that there was some ozone loss after the eruption of Mount Pinatubo, especially in the polar regions, and also some heating of the stratosphere. They found a few other candidates that were more reflective or were easier to handle or had fewer side effects, but none that 'ticked every box' as well as sulphur dioxide. This conclusion has lasted down to today.

Their second task was to design a system for delivering a volcano-sized volume of that gas into the lower stratosphere.

ABANDONED SPICE FIELD TRIAL

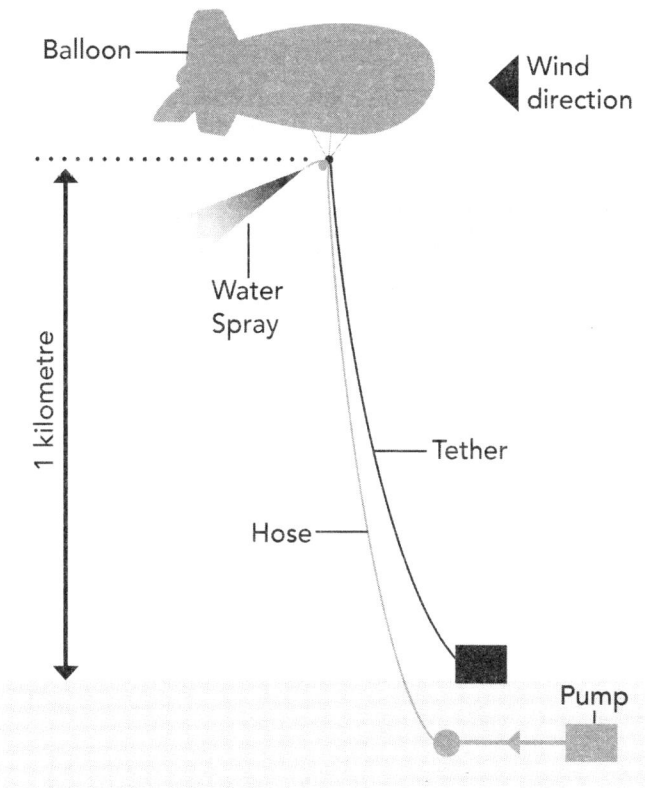

The somewhat bizarre choice fell on a 'tethered balloon – a gigantic balloon connected to the ground by a 23km-long pipe through which the sulphur dioxide would be pumped up into the stratosphere.'* There was no intention to actually build such a system, and indeed there were several major implausibilities in the design, including the lack of any material light and strong enough to carry 23km of its own weight.

The third part was to plug the data into climate models.

> You take information about where the balloon might be, or where one could situate it, and some idea of the materials and the total mass of sulphur dioxide you might use, and see what the total climatological effect was. That produced some very interesting results around what happened if you tried to cool half the planet [e.g., just the Northern Hemisphere]. Nothing particularly good happens... The bottom line is you end up moving rainfall patterns around.
>
> Matt Watson

Such was the state of play then, and in the end the SPICE team didn't even build the one-twentieth scale model that was going to pump 'something like two bathtub-loads of water onto a wet field in Norfolk in October'. What the project did reveal, however, was the nature and the scale of the opposition that any attempt to do real climate engineering would encounter.

> There were definitely people who thought this is opening Pandora's Box, this is the beginning of a dash towards climate engineering... There was some external pressure from Green groups, and it is difficult for me personally

---

* The engineer in the SPICE team happened to be interested in this technology.

because, in general – people like Friends of the Earth and Greenpeace, I'm one of their constituents. I agree with their ethos and to be criticised by them was quite difficult, but I felt that I was right and this was something that needed to be researched.

Matt Watson

The hostility to the whole idea of solar geoengineering in parts of both the scientific community and the environmental movement was evident even before Paul Crutzen's paper broke the silence. Yet advocates of geoengineering research invariably stress that it cannot 'fix' the problem of global warming. They just argue that it might shield us from the most extreme temperature rises for a time and let us preserve a functioning global society while we continue to work on cutting our emissions.

We should think of [geoengineers] as people who can buy us time, give us a bit longer. They're no more a cure for heating than it would be to go on a dialysis machine if your kidneys failed. It merely postpones the moment of your death, that's all. But then of course, who would refuse dialysis if death was the alternative, and we are in roughly that position for geoengineering.

Most of the ideas, like reflecting sunlight back to space by an aerosol in the stratosphere – they would work, all right. So would an aerosol on the ocean surface. They would work, and I think would give us a chance because we'd have longer to get our act together.

James Lovelock

Taking Flight

The sole conclusion about hardware that came from the SPICE programme was that pipes extending 20–30km into the stratosphere borne up by enormous balloons were not a viable technique for delivering SO_2. So what would be? As it happened, at the same time as SPICE (2012) there was a two-man team in the United States looking into exactly this question: Dr Jay Apt, a former astronaut with four Shuttle missions and more than 800 hours in space under his belt, by then a professor of engineering and public policy at Carnegie Mellon University in Pittsburgh, and Dr David Keith, a physicist at Harvard University.

The pair had funding for research into how to get sulphur dioxide into the stratosphere, so they made a deal with Aurora Flight Sciences, a Boeing research subsidiary. The Aurora engineer in charge of the project, Justin McClellan, quickly settled on aircraft as the most cost-effective choice. He then set about designing and a costing a fleet of aircraft that could deliver 1 million, 3 million or 5 million tonnes of sulphur dioxide a year into the stratosphere. (Keith and Apt were not yet certain about the tonnage of SO_2 that would be needed, nor the altitude at which it should be dispersed into the air.)

> The big variables were the quantities, the altitude and some guidance on how long the cruise segment when you're deploying the geoengineering material should be, and how fast you need to be going in order to keep it from clumping up at altitude. We then used an automation tool and designed about 300,000 aircraft for each different amount of mass. Once you go through the full analysis in this automation tool, you end up with about thirty aircraft that can accomplish the mission.
>
> Justin McClellan

The aircraft design McClelland delivered was cheap: for a fleet of 100 planes to deliver one million tonnes of SO_2 a year to the stratosphere, the annual operating cost would be $1–2 billion, equivalent to the operating costs of a small regional airline – and the costs for larger amounts of SO_2 scaled up in an almost linear way. In fact it was shockingly cheap: at least ten countries could afford such a project and have the technological resources to carry it out fairly quickly.

No such planes were ever made, of course. The object of the exercise was to determine whether putting a large amount of an aerosol in the stratosphere was technically and financially feasible, because there was no point in pursuing research into this kind of geoengineering if it wasn't. Once that was settled, the scientists returned to their climate models. More than a decade later no reputable scientist has experimented with delivering even small amounts of sulphur dioxide to the stratosphere, although that may be partly due to the fact that they know small-scale experiments couldn't tell them what they really need to know.

Ulrike Niemeier, a scientist at the Max Planck Institute for Meteorology in Hamburg, has co-authored a number of the key papers on 'stratospheric injection of sulphur dioxide'. She explained why small-scale experiments were unhelpful.

GD: Can't you just test a little bit?

ULRIKE NIEMEIER: No, you can't. At least, not if you want to see a climate impact. If you want to see where the sulphur dioxide is transported, then it might be possible. But assessing the climate impact is impossible because you need a signal that is stronger than climate variability, and climate variability is large. To see an effect in nature we would really need a very strong signal and that's not an experiment. It's deployment.

GD: This leaves us with a parachute that we have not tested.
UN: Yes.

That's why everybody clings to the sulphur dioxide model and doesn't look too far into alternative aerosols in the stratosphere. There's no way to test them, whereas at least Mount Pinatubo has done that experiment for them.

* * *

Working out how to get sulphur dioxide or any alternative aerosol into the stratosphere without a volcanic booster, on the other hand, is quite straightforward. The most recent and most plausible aircraft design comes from Wake Smith, a lecturer on climate engineering at Yale University with thirty years of experience in the commercial aviation industry.

> **WAKE SMITH**: The immediate issue I was after was cost, because you need that to have an intelligent discussion about the cost benefits of this climate intervention versus other climate interventions versus no climate intervention. Having been a financial guy and run an airline, I knew how to build that model. It quickly became clear that existing aircraft can't be modified to reach the required altitude, so I connected with a team of retired Boeing aeronautical design engineers, and with them created the preliminary design for the Strato-spheric Aerosol Injection Loft, or SAIL-One.
>
> What we need is an aerial dump truck that can take a big load of gunk up to the high heavens, dump it there, and come back down for another load. What's unusual is that the altitude it needs to reach – twenty kilometres or 66,000 feet – is roughly twice as high as your Boeing or Airbus airliner cruises. It's the ragged edge of what an air-breathing, fixed-wing, self-propelled aircraft can achieve, but there

are planes that go there – mostly spy planes, but different designs are possible. Our preliminary design is strange to look at: a small, thin fuselage with huge wings hanging off of it, an enormous wing surface area because we need that to stay aloft in the very thin air at 20km up. It's got six engines hanging off those wings, much narrower than the engines on your airliners today because they're low bypass engines, which perform better than modern high bypass engines in this unusual flight regime.

GD: And how much would it cost?

WS: We're in the range of $200 million a copy, after roughly $5 billion development cost. We constrained ourselves in the design of this aircraft entirely to found parts – pre-existing engines, subsystems, wing plan forms, materials – so there's no unobtainium required. The reason this aircraft doesn't exist is that the world hasn't needed a high-altitude dump truck. But it could build one.

GD: How long would it take before one flew?

WS: Assuming a certification path that more nearly mimics a military aircraft than a civilian one, I think five to seven years to get the first airframe in service and then some years thereafter to build the first cohort of aircraft, so let's call it a decade or so.

2035–2135: Imagine...*

Let's imagine that it's 2035, and there has been real progress. A thousand Direct Air Capture hubs and various other CDR initiatives are nibbling at the edges of our annual mountain of CO_2 emissions; renewable and nuclear energies are thriving; and the fossil fuel share of global energy consumption has fallen

---

* CAUTION! This section rests on a series of large and possibly unwarranted assumptions.

dramatically to only 70 per cent. Unfortunately, the remaining CO_2 budget to limit global warming to +1.5°C only had 240 billion tonnes left in it in 2023. Our annual emissions were still around 40 billion tonnes so, despite our best efforts, we ran out of road in 2029.* Nothing apocalyptically dreadful has happened yet, just the steady annual rise in killer heatwaves, hurricanes, wildfires and floods, but the future does not inspire confidence.

By mid-decade, a massive global effort has reduced our annual emissions of 40 billion tonnes of CO_2 per year to only 30 billion tonnes despite continuing economic growth, meaning that we are raising the carbon dioxide in the atmosphere by only 1.5ppm a year instead of the familiar old 2.1ppm. However, even with this change there are 442 parts per million of carbon dioxide in the air, only a little short of the never-exceed 450ppm, and it's clear that we'll reach at least 475ppm before we achieve net zero. Moreover, several tipping points that would involve big further jumps in emissions – methane from thawing permafrost, the shift of the Amazon forest to a savannah, etc. – are starting to activate.

After much discussion and debate the great majority of the world's countries agree to use a Stratospheric Aerosol Injection programme to slow the temperature rise and avoid or at least postpone those tipping points. Fortunately, the implications of implementing such a measure have been rigorously researched since the mid-2020s and have revealed various downsides (as with any atmospheric intervention) but no show-stopping problems. The specialised aircraft needed to make the injections have been under development since 2028, and eight of those planes are ready to go.

In 2035, the first year of operations, 4,000 flights from bases around the world in latitudes 15° and 30° North and South carry

---

* These are not imaginary figures: see Chapter 3.

100,000 metric tonnes of pure sulphur into the stratosphere (each take-off carries 25 tonnes). Up in the stratosphere, the sulphur is mixed with oxygen and released as 200,000 tonnes of sulphur dioxide gas. The goal is to counter the warming effect of *half* of the annual increment of CO_2 in the atmosphere: the maximum we can risk if we are to keep the Intertropical Convergence Zone where it belongs and avoid 'moving the rainfall around'. (For details on why, see Chapter 13.)

In the following year, 2036, it will be necessary to deliver the same amount of sulphur dioxide again, because last year's CO_2 didn't go away. We are just masking half of it, and that's a continuing commitment. Therefore, another six planes and 4,000 flights will be required to counter half the warming effect of this year's increment of CO_2 by spraying an extra 100,000 tonnes of SO_2 into the stratosphere. And so on, adding six more aircraft and an extra 100,000 tonnes of SO_2 each year until 2049. The *effective* extra heating each year will be half of what we would normally have expected from the amount of CO_2 that we dumped into the atmosphere.

By 2050, after fifteen years of operation, there would be almost a hundred aircraft flying 60,000 sorties a year (this might seem a large number, but there are normally this many flights across the North Atlantic every three weeks and these flights would be much shorter). They are delivering 1.5 million tonnes of sulphur dioxide to the atmosphere annually, which is still less than a tenth of the amount that the eruption of Mount Pinatubo put up there.

The cost of this project, averaged over fifteen years of operation, would be just over $3 billion a year (in 2023 dollars). Another 18ppm of carbon dioxide would have been added to the atmosphere (1.5ppm per year over fifteen years), which would certainly violate the 450ppm never-exceed limit, but the *effective* warming in 2050 would be considerably less, as if the world were still at 440ppm. If we have succeeded in cutting our CO_2 emissions further in those fifteen years, say by half to 0.75ppm annually, then we might still be *effectively* at 435ppm (which would feel a good deal safer). Fewer forests will have burned, fewer coastal cities will have been inundated, and the world death toll from overheating (tracked on the daily internet charts that have been standard for all kinds of catastrophes since the Covid pandemic of the early 2020s) will still be below 10 million in most years.

In this scenario, we have been extracting carbon dioxide from the atmosphere in ever larger amounts for twenty-five years, and are finally making inroads into the huge annual burden of new CO_2 emissions despite the doubling of the world economy since 2020. The first contingent of developed countries has, as promised, achieved net-zero by 2050, but China still says it won't be there until 2060 and India says 2070, so nobody questions the need to maintain the SAI programme to counter annual new emissions. We haven't even begun to tackle the trillion-plus tonnes of historic emissions that still fill the atmosphere and the seas – but we have managed to agree that our final objective is to draw down enough of our past emissions to return the atmosphere to the CO_2 concentration of 1990, when we were at around 350ppm.

By 2060, it will be clear that net-zero for India will only be reached, as predicted, in 2070, but emissions elsewhere have been plummeting since the mid-2050s as China nears its own net-zero target. The global average temperature would be at least +2.2°C if it were not for the geoengineering assist, but warming has stabilised at 1.6°C above pre-industrial, so no major feedbacks have been triggered. Various Carbon Dioxide Removal techniques are still working away to reduce the huge accumulation of past emissions (and some will always be needed to counter emissions that we cannot avoid), but by 2090 it should be possible to start reducing the volume of aerosols that we are annually depositing in the atmosphere. A full cessation of SAI might be possible as early as 2135, a century after it began.

And then, once 350ppm has been reached, we will need to have a serious discussion about whether our ultimate target should be lower than that, since after all 350ppm was the level at which the last ice on the planet melted in the previous geological era. Sea level rise happens slowly, but we may prefer to keep the coastlines where they are now.

And that, darlings, is how we saved the world. Lights out now, and sleep tight.

On the other hand...

Perhaps it really could happen like that, but the devil is always in the details, and there are lots of them. There is the possibility that bad things start to happen in the stratosphere after you have been dosing it with sulphur dioxide for five continuous years. Probably not a total disaster: you just stop the process, and after a couple of years all the sulphur dioxide has fallen out of the stratosphere – but now you've lost your get-out-of-jail-free card.

Or you might discover a problem that obliges you to stop doing the sulphur dioxide flights after fifty years, by which time you are holding back a large amount of potential warming because for whatever reason you still haven't got your emissions all the way down. In any case, the carbon dioxide has been piling up in the atmosphere for fifty years, and you have been compensating against the warming that would normally result from 550ppm – only now you get it all back at once. Plants, animals and people all die in huge numbers because they can't deal with such a sudden change. This is called 'termination shock' in the jargon, and it obsesses a lot of people.

Or maybe the problem is political: 'the world' (meaning the major powers) cannot agree on when to start SRM geoengineering, or how to do it, or whether to do it at all, and it never happens. Good luck with +5°C.

Or else the world cannot agree, but some individual country or group of countries decides to go ahead with SRM regardless, and there's a great war, followed by a 'nuclear winter', and those of us who are left can stop worrying about global warming for a while.

Or a new religious creed arises that captures a significant proportion of the world's population and says that all those evil people who are interfering with God's creation must be destroyed, and once again the rest of us forget what the job was and instead waste a decade or two waging another 'war against terror'.

Or perhaps we really do manage to halve our emissions in the next ten years, and halve them again in the following ten, and by then we have abundant fusion power and have forgotten all about those nasty fossil fuels, and everybody lives happily ever after. I'd vote for that one, myself, but I can't seem to find it on the ballot paper.

SRM: Supping with the Devil?

Why, if the risk of a climate catastrophe is so high and SRM techniques are available that might give us more time to deal with it without being overwhelmed by the heating, are so many leading figures in the field set against it?

> [SRM] can be done relatively cheaply. That's part of what makes it deeply dangerous... I desperately want to avoid us

going down that path, because it's just throwing a whole other set of risks into the risk-balancing game. But my deepest reservations about those interventions are human and political reservations, because I see no evidence that we have the collective political maturity to tackle the climate problem in the first place, so the hope that we could get it together on this geoengineering intervention is optimistic at best.

Tim Lenton

It's dangerous nonsense... It would be a complete failure of the human enterprise, because it is supping with the devil. First of all, I'm very doubtful that the governments of the world would agree. The Chinese would have to agree, and the Russians, and we are not able to agree on even the most sensible things. So how to agree on a crazy thing?

The other thing is, we do not really know what the side effects would be. If you talk to David Keith, he will tell you everything is under control. I don't think so, and I have looked into this matter... The very existence of this narrative is a dangerous one. I had a long, long argument with Paul Crutzen about that. It has been discussed many times before – in the Los Alamos lab and so on.

Hans Joachim Schellnhuber, Director Emeritus, Potsdam Institute for Climate Impact Studies

I've interviewed a large portion of the leading scientists in the field, and the objections to SRM that I've heard come in four categories. It seems to me that only one has real substance, but let's go through them one at a time.

The Four Objections to SRM

Objection # 1: The ever-popular 'termination shock' bogeyman
A recent study concluded that only extreme neglect and very bad luck could make a sudden and permanent termination of solar geoengineering necessary. There would never be a greater absolute rise in temperature from abrupt termination than would have occurred if no geoengineering had been done at all, but if we have been holding the heat down for a long time and have not put a serious effort into cutting our emissions, we would be in big trouble if we had to do a sudden stop. About 75 per cent of the 'avoided warming' would reappear within five years, and 85 per cent within ten years.[91]

Suppose it's 2090, the SRM is masking 4°C of warming beyond the 2020 level – and suddenly you have to slam the brakes on hard (never mind why). Emergency termination! What happens next?

Well, obviously, the average global temperature rises by 3 full degrees C in only five years, and billions of people die. Thousands of animal and plant species go extinct. But that's because I set the experiment up that way.

Why did you have to do that crash stop? If you have been basing your aerosol delivery planes at different airfields all around the planet (only sensible economically) and have adequate reserves of aircraft (merely prudent) and good security against terrorist attacks (Standard Operating Procedure), then it's hard to write a plausible scenario where it all suddenly stops and cannot be restarted. And why would you have let your avoided warming build up so much anyway? The whole point of SRM is 'peak-shaving'. It's a temporary measure to prevent extreme temperature rises that might trigger tipping points in the climate system or in human social systems while you work to end your greenhouse gas emissions, not a permanent 'solution' to the warming problem.

> [People say] solar geoengineering is really dangerous because
> if you turn it off suddenly bad things will happen. But in
> every other domain, bad things happening if you stop doing
> something is considered motivation to keep doing it, not
> motivation never to do it. And solar geoengineering being so
> cheap means that you can have all kinds of redundancy. You
> know, five different systems in place, so if one system fails
> another is ready to go. The whole termination shock argument... is a reason why you'd want to build redundancy into
> the system, not a reason not to do it.
>
> Ken Caldeira, Department of Global Ecology, Carnegie
> Institution for Science

Contrary to some of the more cynical propaganda on the topic, once we pursue solar geoengineering (if we ever do), we would not be committed to it forever. We are just strongly advised not to stop it suddenly, especially if it has been going on for many decades and is holding back several degrees of heating. But we are free to gently taper it to a close over a period of time if the world decides it'd rather live with the extra heat that implies. Or, in a more desirable future where we have finally reached zero emissions, we would be free to reduce the aerosol injections in the stratosphere in sync with the CDR techniques that are gradually removing the excess carbon dioxide remaining in the atmosphere. (That's the happy ending.)

Like any other large technological system from national electricity grids and the internet to international air traffic control systems and cold chain delivery systems, SRM systems would require constant management and would be vulnerable to strikes, terrorist attacks and wars, but system-wide outages in such systems are rare and year-long losses of service almost unknown. Moreover, any interruption in a solar geoengineering system of less than a year would be practically without tangible effect, since the average stay-time of the aerosol particles in the stratosphere is estimated to be about a year.

Objection # 2: If people believe that geoengineering is possible, they will stop caring about cutting emissions

Here is the dear old 'moral hazard' argument, made by the wise and prudent but always about the ignorant and feckless, never about themselves. Yet I encounter the ignorant and feckless on a daily basis, and they hardly ever say: 'I can't bring myself to care about my emissions now that I know there's a chance that geoengineering might save us.' The topic just doesn't come up. Besides, the cat is already out of the bag for anybody who has any interest in the subject. Whatever damage it was going to do has already been done, so you might as well get on with it and see if geoengineering *can* save us, or at least contribute to saving us.

> The 'moral hazard' argument is a powerful one... and the folks who make it are doing it in good faith, but my view is that there is far too little understanding that, even if you get down to near zero greenhouse gas emissions by mid-century, the rest of the century is going to be pretty damn uncomfortable unless you do something really serious...
>
> Jay Apt

Objection # 3: The world's governments could never agree on it

Well, yes, it would take an awful shock to bring them around, but the two world wars and the advent of nuclear weapons gave them such an awful shock that they created the United Nations and dozens of other international institutions after 1945, and since then – now over three-quarters of a century – no great powers have fought each other directly. The threat of such a war is ever-present, and the war in Ukraine may have invalidated this statement by the time this book reaches market, but it's a good start. This time we need the great powers to agree on a rather tricky scientific issue *in anticipation of a great shock*, but it seems worth a try, at least. If they don't agree, what have we lost? And as for the risk of their stumbling into a great war, we live with that danger every day over issues as petty as sovereignty over the eastern provinces of Ukraine or the political status of Taiwan. SRM would be just another risk among many.

Objection # 4: The system is too complicated for us to justify meddling in it

Finally, the *complexity problem*, which frightens a lot of knowledgeable people. In recent decades, we have grown our understanding of the climate system and all the other parts of the biosphere that interact with it very fast, but there is still a great deal we don't know. Until we know a lot more, we cannot be certain that when we push the system in *that* direction, it won't instead head off in a different direction, nor can we even be sure that it will return to its normal function if we stop pushing.

> I don't think we're close enough to understanding the complexity, the web, the feedbacks to be able to try some global geoengineering and think we know what the outcome is going to be. A climate state is like a rock at the top of a hill. You can rock it back and forth – which is what we're doing now – but at some point, if you push it, it starts rolling down the hill. If it's a boulder, you're not going to be able to stop

it, and Earth's climate is such a complex system that it has a mind of its own, in some sense, and it is going to find some new state of its own regardless of what you do.

Geoengineering is more dangerous than the problem. You could roll yourself off into some crazy states. We're going to risk tilting the planet into some kind of bizarre global snowball state, not that it would definitely happen, but who knows? We're in completely uncharted territory with geoengineering.

Adam Frank

It seems to me that the rapid warming which will happen if we *don't* slow it down with geoengineering measures is more likely to be the push that starts the rock rolling, but intuition is not a good guide in these matters. The idea that our attempt to slow the warming might roll us into a 'Snowball Earth' seems far-fetched, but there have been such episodes in the past, most recently a pair of 'Snowball Earth' incidents at 717 million and 649 million years ago (around the time when the first multicellular life appeared). On both occasions the ice covered the whole planet all the way to the equator, unlike the partial and rapidly reversing glaciations of the current (but now terminated) Ice Age. Every hypothesis about how and why this happened is controversial, and one of them (but only one) seems potentially relevant to a modest and temporary geoengineering attempt to shave a degree or so off the current anthropogenic warming trend.

That hypothesis is the work of MIT's Constantin Arnscheidt and Daniel Rothman, who proposed in 2020 that Snowball Earths were likely the product of 'rate-induced glaciations'. That is to say the Earth can be tipped into 'Snowball' mode when the level of solar radiation it receives drops quickly over a geologically short period of time. The total amount of solar radiation doesn't have to drop to a particular threshold: as long as the decrease in incoming sunlight occurs faster than a critical rate, a temporary glaciation (aka Snowball Earth) will follow. Specifically, Arnscheidt and Rothman suggest that a 2 per cent drop in the amount of solar radiation reaching the surface that lasts 10,000 years might be enough to bring it on.[92]

If that 2 per cent figure is a bit disturbing, the 10,000-year one is much less so. (The latest proposals for SRM involve blocking about 1 per cent of solar radiation until enough CO_2 has been removed from the atmosphere, which could take one or even two centuries.) In any case, Arnscheidt and Rothman's idea is only one of about half a dozen about the cause of the Snowball Earths, all of them still just hypotheses. Ignorance is dangerous and more research on this and many other aspects of the Earth System is required. But, whether we like it or not, we already know that we will have to make many of our future decisions about preserving a liveable climate from a position of imperfect knowledge.

Given the greenhouse gases that we are loading into the atmosphere, the planet is likely to head off in unpleasant (for us) and to a large extent unforeseeable directions if we do nothing, so we can't make ourselves safe by refusing to act. And we should remember that, while a decision to deploy can be deferred indefinitely, the option of deployment does not even become available until there has been a good deal more research to determine what is safe and what is not. Let's say a decade at least for further research and open-air experiments, with the design and development of specialised aircraft taking place at the same time. Waiting to be sure they will be needed would be a false economy.

Long before 2035, it should be clear whether emissions reductions and Carbon Dioxide Removal techniques alone can keep the parts-per-million low enough to avoid the need for

Solar Radiation Management. If the answer is no (as I strongly suspect it will be), we will want to be able to deploy SRM right away. And as a young Finnish climate activist pointed out to me, there's another reason why SRM should not be left languishing in the last-chance ghetto.

> People don't even want to think about geoengineering. The discourse that has been developing for two or three decades is that it's Plan B: 'Okay, geoengineering is a crazy idea, we should look into it only when it's too late.' And since most people don't want to accept that it's too late, they shun geoengineering. We really try to de-construct this false narrative, because so long as it stands we are not going to consider geoengineering seriously at all.
>
> **Anton Keskinen, co-founder of Operatio Arctis, Helsinki**

When it comes to climate, last-minute interventions are virtually guaranteed not to work. There is no Plan B. By the time we realise we're hitting a really big tipping point, it will generally be too late to break the momentum of the system.

Commitment to full-scale research and development of SRM, including aircraft design, should come no later than 2026 or 2027, with a view to being ready for a decision on deployment by 2035. I have little faith that such a deadline will be met, of course, but every year of delay past that date will increase the risk that we have no viable fall-back options if (or when) our current highly optimistic projections fail to come true.

> I've tried to be so careful for so long when talking about solar geoengineering... And yet here's where I get myself into trouble with some of my colleagues... If we want to live in the world as it is, and not the world as we wish it could be, then

solar geoengineering is the sort of thing which I think needs to be given serious consideration. And so I agree with you, I think the timeline you're setting out is about right. [I had suggested that most people will want relief from climate-related catastrophes badly enough to favour solar geoengineering by 2035.]

What we're seeing is climate impacts happening far more rapidly with far more devastating effects than had been forecast. We didn't think that the Arctic would crash by now, and yet it's almost gone. We didn't think we'd be seeing these wildfires in Australia and the United States and elsewhere with the frequency and severity that they are being seen. Given that we're at about one degree Celsius, we thought those were far-distant prospects, and so 1.5 degrees of warming above pre-industrial averages could turn out to be far more devastating than had been imagined when that target was set as the threshold for international action.

As people start to experience these climate impacts, one has to imagine or at least hope that there's going to be more push for action, at a time when actions that will have a near-term impact on the climate system are going to be limited. About the only thing that can very quickly bring down the temperatures is solar engineering. Mitigation activities and all the carbon removal activities will take a long time to yield the sort of reductions that are our ultimate goal, so one can imagine that by this 2035 target you've set, the agitation for research into or potential deployment of solar geoengineering will be growing.

Simon Nicholson, Forum for Climate Engineering
Assessment, American University

13

RISKS... AND A RECOMMENDATION

I can smell my bridges burning as I start this chapter – bridges of respect, of trust, and in some cases even of friendship with the scientists who shared their time with me while I was researching *Intervention Earth*. They are working hard, their science is sound and indispensable, and their intentions are good. Without their research, we would be doomed (and I am not speaking rhetorically). But I do not think that the policy recommendations many of them would make are really based on their scientific findings. Instead, at the last moment, they draw back in something like horror from the implications of their own conclusions.

When I was researching my first book on this subject a dozen years ago, I ended it by saying that we didn't have to settle the debate about geoengineering right away. 'Let's wait five or ten years, and if those "early and steep cuts in emissions" still haven't happened, then we'll discuss it again. There's time enough for that. In the meantime, though, I'd like lots of research to be done on geoengineering techniques of all kinds, because I suspect we will need them.'

Well, many years have passed, and there still haven't been any net cuts in global greenhouse gas emissions. More research has been done on solar geoengineering, but not one tenth of what could and should have been done. The money wasn't there, of course, but what made the money scarce was probably the same thing that made so many of the scientists I interviewed reluctant to discuss geoengineering at all. They would offer a brief list of reasons why it was too dangerous and wouldn't work anyway, and move on quickly to a more comfortable subject.

Generally I didn't argue with them, because many of the reasons they cited for opposing geoengineering were arguments that had already been refuted by the work of the relatively small number of scientists who are working in the field, and I think many of the naysayers knew that. They just didn't want to get into a discussion on a subject that was a trigger-point for some of their angrier colleagues. Paul Crutzen broke the absolute taboo on discussing geoengineering in 2006, but there is still a chill on the subject that deters most people in the climate science world from engaging with it. Here are a couple of examples of how they deal with – or rather, don't deal with – the matter. I have left the speakers anonymous, but I have heard these replies many times from people who know how dangerous the situation is.

> Solar Radiation Modification is basically trying to buy time. That's an argument that's often used, but at the same time we continue emitting CO_2, so we make the problems worse. We put up a sunscreen which tries to compensate for some symptoms, but it doesn't get at the cause of the problem and I think that's needed. That's ethically irresponsible, promising to compensate for symptoms for much longer than my lifetime.

This false equation appears in some form in almost every criticism of solar geoengineering: the notion that if we deploy SRM while continuing to burn carbon, we are making the problems worse. I think this is 'moral hazard' in disguise. Every advocate of research on SRM always stresses that it must be accompanied by a maximum effort on reducing emissions. The sole reason for deploying SRM would be that we still haven't managed to bring our CO_2 emissions down enough, the heat is rising fast, and we are at risk of activating irreversible tipping elements that would cause such damage to society that further efforts at solving the problem by other means might be abandoned.

> We can't guarantee that SRM can be done continuously, and it would have to be done for many decades, if not centuries, irrespective of what government you have, what other problems you have, and what weather conditions you have. Mankind has never managed to have anything stable for several decades, let alone a century. I don't think this is a responsible way of acting.

Yes, it would have to be done for many decades, and perhaps centuries. *Everything* about dealing with climate change operates on a scale of decades or centuries. There is no 'solution' around the corner after which we can celebrate our success and get back to our previous lives. Coping with climate change by some combination of emissions cutting, geoengineering and adaptation will be a primary concern of people everywhere for at least the next several generations. If our attention span as a species is not up to that standard, then we will end up among the many intelligent civilisations in Adam Frank's virtual (but very Darwinian) universe that just don't make the grade.

This outcome is not inevitable, because our emerging global civilisation has demonstrated that it can stay focused on complex international tasks for much longer than a decade or two.

The most dramatic example is nuclear weapons, which were used in war within weeks of being created in 1945 – and have not been used in war again for more than three quarters of a century. It took relentless attention and grudging cooperation (often unacknowledged) between a wide variety of real and potential enemies to make this happen in a world of many independent states where war remains possible and governments still prepare for it. Enough people recognised that avoiding nuclear war was a matter of survival and have acted accordingly over several generations. Now we are coming to realise that runaway global heating is a comparable threat to our collective survival, and there is no reason why we cannot provide a similarly grown-up response to that. There's just no guarantee, that's all.

SAI: the Complexity Problem

There are two potentially serious technical objections to Stratospheric Aerosol Injection (SAI). One is the 'complexity problem', which could be a deal-killer if it really makes managing any major intervention in the climate impossibly dangerous, as its proponents claim. There is also a major political problem to sort out, which is how countries can agree the rules for spraying aerosols into the stratosphere. This 'governance' question turns out to be closely linked to the 'complexity' issue, however, so let's address complexity first.

Up until 2016–17, Solar Radiation Management was a pretty primitive business. In particular, Stratospheric Aerosol Injection, the main hope for cooling the climate, had all the sophistication of a sledgehammer. The idea was to boost a large amount of sulphur dioxide into the air, probably by specially designed aircraft, and spray it into the lower stratosphere more or less above the equator so that the upper atmosphere winds distributed it fairly evenly all the way around both the northern and southern hemispheres. The amount of sulphur dioxide you would put up was a modest fraction of what a big volcano like Mount Pinatubo produced, so you knew you were operating within safe bounds. Then you set your experiment running and watched what happened next.

You aimed for an overall 'solar dimming' that would bring down the average global temperature by perhaps half a degree

Celsius. But what did that do to the Indian summer monsoon? To the melting of the Greenland ice cap? To wildfires in western North America? Who would win from these changes, and who would lose, and what were they likely to do about it?

You didn't do this experiment in real life, of course, or they'd have come for you with flaming torches and pitchforks. You just loaded it into your climate models, the same way you would do for any other question. Put it into enough different climate models, and if almost all of them spit out similar answers then you might begin to trust them. But SAI in its early form was clearly a very blunt instrument, and persuading people to try it – even people who were suffering huge damage from increasingly wild weather, let alone those who stood to lose from it – would have been a very hard sell.

Rain and its Absence

In the early days of SAI they ran into an unexpected problem. More heat causes more evaporation and therefore more rainfall – but when you use sulfates in the stratosphere to bring the heat back down, the rainfall doesn't just go back down at the same rate. It drops beyond its previous level and causes artificial droughts, especially around the equator. The fact that 'solar dimming' would produce 'winners' and 'losers' was seen as a fatal flaw by many critics: the victims would object, and a global consensus for deployment could never be attained

Even today you will still hear this 'global drying' issue cited by people who haven't been paying attention, but what the scientists who were interested in geoengineering did, with Peter Irvine of University College London and David Keith of Harvard University in the lead, was to investigate further and come up with a quite simple solution. If allowing, say, two degrees of warming to happen without intervening by geoengineering will cause more precipitation than you want, and countering the full two degrees by geoengineering will cause unacceptable droughts, what would countering half the warming do? Plug that into the same models as before, and the answer, to nobody's vast surprise, is that global rainfall stays where it was before you started geoengineering. Okay, so you're only offsetting half the warming, but that's a great deal better than nothing.[93]

These were the Dark Ages of SRM, and they're nothing to be ashamed of. Every really good idea starts simple, and grows more complex as it is applied in the real world – or, in this case, in a mathematical model of the real world. The catalyst that freed solar geoengineering from the clunky 'solar dimming' approach was an encounter between Ben Kravitz, a PhD student working on geoengineering topics with Ken Caldeira at the Carnegie Institution for Science in Palo Alto, California, and an aeronautical and aerospace engineer called Doug MacMartin who had a strong interest in climate engineering techniques.

We always used to think; just put the sulphur up there over the equator and let the planet do the work for us, because the planet's job is to take stuff from the tropics and move it onwards. Then we started playing around with the research on volcanic eruptions and saw that there are different effects depending on where the volcano exploded. So I basically cornered Doug MacMartin and said: 'Teach me control theory'. Then, at a certain point, we got away from 'solar dimming'. This was like 2016, 2017, and we turned the question around. Not 'if people did geoengineering, what would happen?' That's the wrong question. The geoengineering will do whatever you tell it to do, within certain bounds. We started asking: 'What can you tell it to do?'

Ben Kravitz, Department of Earth and Atmospheric Sciences, Indiana University

We pause here for a brief excursion into control theory. Doug MacMartin knows all about showers:

> How do you design algorithms for things; a self-driving car or an autopilot for an aeroplane? It's not fundamentally different from what you would do if you were taking a shower. Temperature too warm, you turn the knob one way. Temperature too cool, you turn the knob the other way. You don't need to know everything perfectly. If you don't quite know how much aerosols to put in to get a given amount of cooling, well, you'll find out. You'll just adjust it, even if the details of the algorithm are a little bit unclear.

> With stratospheric aerosols, the time constraints on the aerosols in the atmosphere are on the order of a year, which means that you can't manage things that are faster than that. A lot of people confuse geoengineering with an attempt to manage the weather, which is inherently chaotic – the butterfly flapping its wings over Beijing and all that – and we're nowhere near close to the ability to tweak the weather. But the time-scales we're talking about with SAI are years, because those are the time-scales of the climate system.

This, believe it or not, is where the 'complexity problem' comes from: a confusion, sometimes deliberate but mostly not, between the insane and impossible complexity of trying to control the weather across an entire planet on an hourly, daily or weekly basis, and, on the other hand, the moderately complicated but quite slow-moving job of tweaking the average climate over large regions. Climate, not weather. The climate, by definition, is entirely and exclusively about averages. Tweaking it is still a demanding task that requires a lot more research before anybody tries to do it in real time, but the 'complexity problem' does not exist. It is a category error.[94]

> **DOUG MACMARTIN:** By the time you average over a year or more and you average over the scale of the whole planet, you've averaged out all that sort of variability and you're left with more or less manageable problems. Like I said, it's not that different from what you would do in a shower; the slight difference is that you might be adjusting more than one knob, so to speak. You're not just saying 'I want to make sure that the global mean temperature is within some range', but you might also say 'I want to make sure the Northern Hemisphere versus the Southern Hemisphere difference stays roughly constant because that affects precipitation patterns in the tropics'.
>
> **GD:** But you can twiddle both those knobs?
>
> **DM:** You can twiddle both of those knobs and you can do them more or less independently.

Not only that, but there are more than two knobs. You can also vary where you inject the aerosols in terms of latitude, volume, altitude, season, etc. Kravitz and MacMartin talk in terms of 'degrees of freedom', but we can think of them as knobs, and each can control for a different aspect of the climate or even a certain latitude. They haven't finished exploring this in their models yet, but MacMartin and Kravitz think there are probably at least six to eight knobs – different settings for different functions of the climate that you could 'programme' at the same time.

They are pretty broad settings – no fine tuning is possible – but this approach transformed the work of the geoengineering team at the National Centre for Atmospheric Research (NCAR) in Boulder, Colorado. They applied this 'feedback control algorithm' to their models and came up with the answer that the best 'injection sites' for the SAI would be at 15 and 30 degrees both north and south of the equator – but the algorithm gave them a lot more flexibility than that.

> Magnitude is the easy one: you put more [sulphur dioxide] up, you cool the planet more. Altitude is pretty similar: the higher up you put stuff, the longer it takes to fall down, so you get more bang for your buck. The most interesting ones to me are latitude and time of year...
>
> Latitude really is the big knob: you get different effects if you put it at high latitudes around the poles versus mid-latitudes versus the tropics. If you put it up in the Northern Hemisphere, you mostly cool the Northern Hemisphere; vice-versa for the Southern Hemisphere. If you put it up in the tropics, you get both.
>
> [As for] time of year... you don't have to put it up all year round if you're trying to cool the Arctic. You don't inject in the winter in the Arctic: there's no sunlight, so it's not reflecting anything. And different seasons can also have different effects for things that have a seasonal character, like the Indian monsoon. In the same model simulations, the opposite season is better for the Amazon, so you can start to see where the trade-offs emerge.
>
> Ben Kravitz

If you want to cancel more of the warming and you have run out of knobs, you might consider the possibility of spectrally tuning the aerosol you are putting into the stratosphere to the near-infrared wavelengths. This might allow you to maintain rainfall at the pre-geoengineering level around the world, even as you cool the mean global temperature increase by more than half.[95] There is also a potential opportunity to fine-tune the process by coordinating with marine cloud brightening efforts.

All of this is still tentative stuff, but it has opened up new horizons for the geoengineers. Maybe they could use their knobs to create a climate that resembles the recent (but considerably gentler) past, leaves no sore losers and keeps us away from any possible 'tipping points'. They call it a 'design space' (nobody say 'Pandora's Box'), and in the past five years there has been a great leap forward in the science of geoengineering. More targeted, specific interventions in the atmosphere are becoming possible, but we are still a long way away from what you might call precision.

Stratospheric aerosol offers you the ability to control the climate, which sounds quite scary, and you might imagine that we could steer a hurricane so that a big weather event happens or doesn't happen. That is not the level of control we would have. It would be much coarser. The aerosols can only be controlled over the course of a few years, like do we want the aerosol layer to thin over the next few years or thicken a little bit? And you don't get to control the exact spatial distribution; it's quite coarse, broad in its application. You've got a choice between a few different distributions of aerosols and a range of climate outcomes sort of under your control.

Dr Peter Irvine, Department of Earth Sciences,
University College London

SRM is not a sleek getaway vehicle that will let us leave the drudgery of emissions cuts and carbon dioxide removal behind. It's relatively cheap and fast-acting, but it cannot set us free; it can only buy us more time for the long haul and the heavy lifting to happen. Since there are no plausible alternatives for buying that time, we should count ourselves fortunate: SAI has been pre-tested by large volcanic explosions that have put sulphur into the stratosphere every fifty years or so, and the consequences of doing something similar ourselves do not seem particularly hazardous. In fact, they seem mostly benign. David Keith has been working on geoengineering for about thirty years, and is a leader in both Solar Radiation Management and Carbon Dioxide Removal. The experiment he describes here did not even benefit from the sort of 'design space' manipulation of the inputs that has been discussed above.

I and many others twenty years ago, said: 'Solar geo can manage global temperatures, but it's going to make some climate variables a lot worse.' So I started a collaboration with the Geophysical Fluid Dynamics Laboratory at Princeton, one of the oldest modelling labs in the world. They have a model that is in some respects one of the best in the world, especially for tropical cyclones. They never worked on this topic before, so it was a 'clean' experiment. They had their own people setting it up and running it, and we analysed for some key variables like availability of water, peak precipitation and peak temperatures – and whether or not some SRM tended to make things better or worse (that is closer to or farther from pre-industrial conditions). The result was almost too good to be true.

I feel like a salesman, even though I'm just telling you what the model said. Almost all regions show water availability moving back towards pre-industrial. That's true for peak precipitation as well, and certainly for temperature (which we already knew). My sense is that the result is too good to be true. The real world will not turn out to be that good. But it is striking that when you choose a best-in-class climate model and try it cleanly for the first time, you get a result like that. It's not necessarily a reason to deploy solar geoengineering, but the fact that over very large tracts of the world most climate hazards would be reduced is a reason to take it very seriously.

David Keith, Gordon McKay Professor of Applied Physics, Paulson School of Engineering and Applied Sciences, Harvard University

It's almost certain that nobody will deploy SAI or any other SRM technique in this decade. Even the scientists themselves don't feel ready to offer it to the public. As Doug MacMartin put it: 'How would you convince people? The answer would be you do the research, but there's vastly more research to do than there is time, particularly given current research budgets. So you would really need to ramp up research, or else accept an enormous amount of uncertainty.' There's always a chance that some miracle might yet occur that motivates us to make 45 per cent cuts in CO_2 by 2030, or at least somewhere close to that, so why not wait until the evidence is in? The snag is we need that evidence before we decide – and we won't have it. In MacMartin's words: 'We're not currently on a pathway to deliver in the ten or fifteen-year time frame you mentioned.' .

Soot

If the unspoken Plan B in many people's minds is a rapid resort to solar geoengineering if tipping points begin to cascade, the hidden catch has always been that SAI can't be deployed rapidly. The planes that could deliver sulphur dioxide aerosols to the stratosphere don't exist, and even in an emergency it would probably take seven or eight years to design, build, test and mass-produce them. If you think that there may well be a crisis in the mid-2030s, you would want to see that development process well underway by the late 2020s – and that is not likely. Procrastination has been the defining feature of the response to the climate problem, and there is no reason to expect that to change until the crisis is a good deal worse.

It's therefore somewhat comforting to learn that scientists in Boulder working on geoengineering at NCAR (National Center for Atmospheric Research) have come up with a potential way to get sulphur dioxide into the stratosphere without the aid of specially designed ultra-high-altitude aircraft.

Here's an extract from the abstract of the relevant scientific paper, which bears the cheery title 'Toward practical stratospheric aerosol albedo modification: Solar-Powered Lofting'.

Stratospheric aerosol injection involves introducing large amounts of (sulphur dioxide) well within the stratosphere... thereby increasing reflection of solar radiation. We explore a delivery method termed Solar-Powered Lofting (SPL) that uses solar energy to loft (sulphates) injected at lower altitudes accessible by conventional aircraft.

[Black carbon] particles that absorb solar radiation are dispersed with the sulphur dioxide and heat the surrounding air. The heated air rises, carrying the (sulphates) to the stratosphere. Global model simulations show that black carbon aerosol (10 micrograms per cubic metre) is sufficient to quickly loft sulphur dioxide well into the stratosphere. Solar-Powered Lofting could make SAI viable at present, is also more energy efficient, and disperses sulphur dioxide faster than direct stratospheric injection.[96]

Simone Tilmes, head of the NCAR team, summarises:

> The difference here is that you don't need to reach up to those higher altitudes and you let nature, the sun, do the work for you. It can be a game-changer, if it works...

If it works, it would mean that there's no need for a long wait for new-design aircraft able to operate at 65,000ft, the altitude of the lower stratosphere in the tropics. (Near the poles it is significantly lower.) If you're suddenly in a great hurry to geo-engineer your way out of a climate emergency, just tear the seats out of some Gulfstream 650 business jets (good for 50,000ft), install tanks and spray equipment instead, and let the sun do the rest of the work.

As an added benefit, black carbon, a major warming source at lower altitudes, operates as a cooling element in the stratosphere. It's really soot, of course, which isn't ideal in any part of the atmosphere, but it will eventually fall out again, it isn't in the meantime harming people and animals (there are no living things in the stratosphere), and it gives you a quick route to the sunshade you wanted.

There is little to love about this particular pact with the devil: using black carbon to loft sulphur dioxide into the stratosphere is clearly part of the slow but inexorable process of becoming 'planetary maintenance engineers' that Lovelock warned us against forty-odd years ago. But we have left it much too late to turn back the clock by gentle, non-intrusive, 'natural' methods. Either we try to clean up our own mess, using harsh chemicals when necessary, or we accept that we and most of the animal and plant species we share this planet with must endure a severe and prolonged global heating episode involving massive die-backs and many extinctions.

Ozone Worries

There is one particular area of research that must provide hard answers before any serious commitment is made to SAI: how much damage would sulphates in the stratosphere do to the ozone layer? The rapid and comprehensive ban on CFCs under the Montreal Protocol of 1987 had stabilised ozone levels over the poles by the 2010s, with a reasonable prospect that even these long-lived chemicals would have decayed enough by the 2060s to allow a return to the pre-CFC ozone concentration and a closing of the 'ozone holes'.

Paul Crutzen was aware that chlorine gas from CFCs was more effective in destroying ozone in the stratosphere when sulphates were present, but he reckoned that by the time solar geoengineering was implemented, the CFCs would largely be gone from the atmosphere. The Nobel Prize for his work on ozone gave his opinion great weight, but subsequent changes in the climate have unfolded more rapidly than he and most other people expected. Deployment of SAI in the 2030s or 2040s, when CFCs may still be around in the stratosphere in significant quantities, is now a possibility, so we must re-assess the interaction of sulphates, CFCs and ozone destruction.

This is particularly worrisome because scientists observed that after the devastating bush-fires in Australia in 2019–20, the ozone hole over Antarctica expanded by 10 per cent. By 2022 they had concluded that organic acids in the smoke from the wildfires had combined with the chlorine compounds in the stratosphere from CFCs to produce chlorine monoxide – 'the ultimate ozone-depleting molecule'.[97]

Unless the global temperature stabilises or drops, there will be lots more wildfires in Australia, South Africa, Argentina, etc., and the southern ozone hole may start expanding again, allowing more ultraviolet radiation to reach the ground and causing more blindness, skin cancers, etc. in the Southern Hemisphere. (Maybe the much smaller northern ozone hole will also start to grow: there are now plenty of massive wildfires in North America, Europe and northern Asia too.) But the worst effect could be to make 'Stratospheric Aerosol Injection' unacceptable because of its impact on the ozone holes.

Some trade-offs are inevitable in any intervention in the climate, but the prospect of rapidly expanding ozone holes would require at best a search for an alternative non-sulphate aerosol, at worst the abandonment of the whole idea of Stratospheric Aerosol Injection. If that is taken off the table, the prospects for the future darken considerably.

Marine Cloud Brightening

And so to the kinder, gentler kind of SRM: Marine Cloud Brightening (MCB). Perhaps because it's viewed as a merely 'local' technique – spray some seawater, thicken up some low-lying clouds, reflect a bit more sunlight – MCB doesn't attract the fear and anger that Stratospheric Aerosol Injection awakens in some people. Instead, it's treated with condescension, or simply ignored.

Marine Cloud Brightening came from the observations of Irish meteorologist Sean Twomey, who realised that the smaller the average size of droplets in a cloud, the more sunlight it reflected. John Latham took that notion (the 'Twomey effect') and hypothesised that it should be possible to make clouds over the ocean reflect more incoming sunlight by spraying tiny particles of salt into the clouds to serve as condensation nuclei.

The size of the droplets, though, was critical, as Alan Gadian of Leeds University explains:

> Think of the weather as a system for transporting heat from a hot equator to a cold pole... My argument is that you want to prevent the heat from reaching the poles, and that's why I suggest we cool the belts of marine stratocumulus clouds that cover perhaps 5–10 per cent of the globe. One off Namibia, one off South America, one off North America.
>
> If I could change the droplet sizes by a very small amount (1–5 per cent) in those specific areas – just 5 per cent of the globe – I could easily give you global warming or global cooling of 3.5 Watts per square metre.
>
> Alan Gadian, senior research scientist, Leeds University

The '3.5 Watts per square metre' mentioned by Gadian would equate to doubling or halving the amount of CO_2 in the atmosphere, and would raise or lower the average global temperature by 2°C. But how could you do such a thing in practice? John Latham envisaged unmanned, remotely controlled vessels that would spray tiny droplets of purified seawater into the air near the sea surface ('the marine boundary layer') in areas where marine stratocumulus clouds are plentiful. Most of the droplets would evaporate, leaving tiny grains of salt that would be wafted up into the cloud by convection. Those granules of salt would recombine with moisture in the clouds to form tiny, highly reflective droplets of water – which would, because they were so small and numerous, reflect much more sunlight than ordinary droplets. The only tricky bit – well, one of the tricky bits – was that he needed ship-borne sprayers that would churn out ten trillion 800-nanometer water droplets per second from billions of sub-micron nozzles etched into silicon wafers.

Latham first outlined the idea in an article in *Nature* in 1990, but it received little attention because not many people were concerned about global heating at that time. He kept working away at it, however, and by 2004 he knew he was on to something important. Stephen Salter, Emeritus Professor of Engineering Design at the University of Edinburgh, explains:

> **STEPHEN SALTER:** John was very surprised to find out how little water you'd need to spray to do the job. It works out that we could offset all the damage we've done to the climate up to now if we could spray about ten cubic metres of water a second of the right droplet size in the right place at the right time of year. He asked me if I could make the spray that he would want. I said: 'Yes, I can do that' – which was not totally accurate, but I've been working on it since about 2004, and I think we are very nearly there.
>
> We've done all the engineering calculations and drawings. What we need now is to start making some laboratory tests to see whether the ideas we've had for making these tiny drops are going to work. Up to now we've had an incredibly high progress-to-cost ratio. It's infinite.
>
> **GD:** No money?
>
> **SS:** No money at all, but quite a lot of progress.

John Latham died in 2021, and today there is still little funding for computer time, no hardware to test, nothing really except the free labour and limited financial resources of a few dedicated individuals. Stephen Salter is now eighty-two, but he continues to work on it every day. There are some younger academics at universities in Israel, Norway, Australia, the UK and the US working part-time on various aspects of the concept, mainly the physics that happens within the clouds. The Silver-Lining 'non-profit' is very secretive about its activities but is subsidising a few American climate scientists on it.

The sole scientific team actually doing outdoor research with aerosol-generating equipment is Daniel Harrison's group at Southern Cross University in Lismore, New South Wales, which began looking into the possibility of using MCB to protect the Great Barrier Reef after the mass bleaching event of 2016. Their first seagoing tests were in February 2020, and it now appears likely that a significant proportion of the aerosols do actually make their way up into the clouds, where they increase the proportion of very small droplets and raise the clouds' reflectivity by around 10 per cent. Because their work is restricted to that single 2,500-km reef (and they have the support of the Australian government), they have avoided the obstacles that normally hinder work on any kind of SRM in other jurisdictions, and their studies continue. But that's it: the total number of climate scientists working anything near to full-time on MCB probably doesn't exceed a dozen.

There is one paper by some people in Norway suggesting that we can do quite a good job in clear skies, but the jury's still a bit out on that. I'd like to see replication of that work on clear sky. The advantage of working under a clear sky is that the air, especially if it has just rained, will be very clean, and the Twomey effect works best where the air is clean in mid-ocean. It doesn't work over land because there's already so much aerosol in the air that you can't really make a difference.

We think that if you could put them in the right place at the right time, you'd need about 300 of the ships we've designed. The ships would each cost between US $3–4 million. Multiply that by 300, and maybe double it again because all the ships won't always be in the right places, and the initial cost is $1.8–2.4 billion.

Stephen Salter

Deployment is a different question for Marine Cloud Brightening (MCB) than for Stratospheric Aerosol Injection, because it doesn't have to be planet-wide. It can't be; it only works over the oceans – and although some form of international regulation or insurance would be needed to deal with any damage from unintended side effects, the fact that MCB operations could be moved to another part of the ocean or simply switched off immediately if they were causing a problem makes deployment issues a lot more tractable. Financing cloud-brightening operations could also be more straightforward than SAI, which offers no obvious commercial return to an investor and therefore must depend on cooperating governments to subsidise it. The benefits of MCB could be relatively local, and therefore more saleable.

SS: If you got a group of governments from, let's say, the Gulf of Mexico who are worried about the damage from hurricanes… What you could say is that we will get the sea surface temperature down from 27°C to 24°C, and you pay us on how close we got to 24°C. We will have these things cruising around spraying between Africa and the Gulf of Mexico and we'll be measuring the sea surface temperature. At the end of the hurricane season, you can decide how much of the fee we've actually earned. Now that would be a way that could look attractive either to governments or maybe to the insurance companies who are insuring against hurricane damage.

GD: You would do the cooling on the African side of the Atlantic, where most of the hurricanes form, and, even if they do form, they will be far less powerful?

SS: Yes, and that looks to be the most commercially attractive route.

The Governance Issue

So much for the many complexities of SRM, but all of this is just talk unless geoengineering gets around all the obstacles and out into the real world. And that's where the 'governance' issue looms large.

Many people believe the opposition to SRM is so strong that it can never be deployed, but most of them seem to be weighing up the apparent risks of geoengineering against the apparent security and stability of the climatic status quo. That is not the context in which the decision will be made. It will be made in a world where damage from increasingly wild weather has put its thumb firmly on the scale in favour of deployment, but where the familiar geopolitical concerns and rivalries have not vanished or even receded. Joshua Horton is the research director for geoengineering policy and governance issues with the David Keith Group at Harvard University.

JOSHUA HORTON: If geoengineering ever materialises, it's going to enter a world that's full of very big things going on, and it will not be number one in terms of national priorities. It'll be important, but it's going to come after other things... Obviously, the key dynamic at play in the world right now is the growing rivalry between the US and China, and if you look at the details of how it's unfolding, one of the key axes is about status and prestige.

This is not to say that material interests don't drive things too. Of course they do. Security concerns, worries about prosperity, ideological issues, authoritarianism and democracy, this kind of stuff. But one thread that cuts through a lot of what's going on is the great importance of status and prestige. And if you think about how geoengineering might end up being received in the middle of this – my hunch is that it's going to arrive as a status issue.

Geoengineering has no real commercial potential. It's got no real military potential either; technically speaking… That leaves you looking at the possibility that like a lot of big science projects or military or infrastructure projects… prestige will be the lens through which these two countries and others look at what to do with geoengineering. It's a pretty wild idea, and building this sort of planetary mega-project, this global infrastructure, is associated with prestige and status. I think it's going to be hard for both parties not to look at this as an opportunity to gain prestige, enhance global status, [posture] as the world's saviour or something. It's a little crazy, but that's how it's going to be received.

GD: That's relatively promising because it makes non-zero-sum games possible.

JH: Yes… issues of prestige and status tend to evolve in one of two directions. One is that there's a claim to be number one in a hierarchy, and then there's no sharing. 'Who's top dog?' If that's the way geoengineering comes to be perceived by either one of these countries, then we've got a problem. But the other way that prestige projects and status endeavours can evolve is through sort of a 'club' approach, like the G7. A select few countries, by virtue of membership in this group, would achieve an elevated status – and it's not at each other's expense. It's not zero-sum. It can be win-win, which is a popular phrase among Chinese diplomats.

To those who never see actual diplomats at work, this must sound hopelessly naive. But cooperation does happen, bargains are made and kept, and rational analysis sounds much the same in every language.

Civilisation is not a howling wilderness of dog-eat-dog savagery; it is an imperfectly rational mosaic of societies that usually manages to create and observe rules that make domestic and international cooperation possible. Sometimes it fails, and sometimes tragedies happen as a result, but collaboration is the likelier outcome when more of every country's interests are aligned than opposed. That is the case with the threat of global heating, so David Keith assumes that the existing limited cooperation in the domain of climate change will continue and may even grow, despite setbacks and local failures.

You often hear people say that geoengineering will only work if we have unanimous agreement. That's completely naïve. Nothing that we do in the world is by unanimous agreement. Forget it. The other thing people say is that some single billionaire will do this. That's ridiculous. There are all sorts of checks and balances, and it would be stopped.

I'd say that if both China and the US don't want solar geoengineering, it won't happen. If they both do want it, it will happen. The interesting cases are if neither of them has a clear answer because of their internal politics. Then smaller countries or coalitions matter more. One way this works is that some major social and environmental organisations come to advocate for it, and then a coalition mixing countries from the Global South and the Global North advances a coherent plan for it. That might be quite a small group of countries at first, and then a bunch of other countries, hopefully most, are willing to go along with that plan, leaving a relatively small number opposed. For me, that's also a plausible way for it to happen.

David Keith

Extreme Weather Insurance

If solar geoengineering does happen, the biggest political problem is going to be compensation. Even if the geoengineers come up with such an elegant and inclusive design for injecting the sulphate aerosols into the best locations at the right times that every major region enjoys a reasonably accurate replica of its pre-warming climate, there are bound to be some places and times when wild weather does serious damage to crops, property and lives. Natural climate variability will continue, including rare but extreme events: 'You must expect that in any given year 1 per cent of the world's population will experience a once-in-a-century flood,' as Doug MacMartin put it.

There will continue to be heatwaves, floods and storm damage. If geoengineering is underway, then people will inevitably blame these events on those who are funding the geoengineering, and there's no way of proving whether they are right or wrong. So would some equivalent of state-sponsored 'no-fault' car insurance be a useful precedent for extreme weather damage?

> The idea of spending a lot of time in court, not only is it insane, nobody wants it, but it's also a practical impossibility because liability in tort law just doesn't translate to these kinds of damage claims. That's a non-starter globally for a million reasons. So we've discussed an emerging alternative approach to providing compensation for climatic harms. It isn't perfect by any stretch, but it seems to be promising.
>
> It's this idea of 'parametric' or index-based insurance,

which wouldn't rely on demonstrating that climate change or geoengineering or any sort of climate intervention or phenomenon caused or didn't cause severe damage. It's simply based upon whether a certain amount of damage happened, based upon thresholds that are written down in agreements in terms of temperature thresholds, or if it's about hurricanes then wind-speed maximums, or if it's flooding then the amount of rainfall over a five-day period. These thresholds are set by expectations of what climate would normally do [in a warming climate] irrespective of what geoengineering might cause to happen, and agreed to before any contestation.

Joshua Horton

Then, if all the roofs in your town are torn off by an out-of-season Force 5 hurricane, or wildfires devastate more than 500 square km of your territory, there's an automatic pay-out, no questions asked. Every country involved in the geoengineering project would have to sign up for this insurance, and countries not involved would be offered a reduced rate to encourage them to participate. This could become very expensive if the weather gets more violent once you start geoengineering, but the project is based on the assumption that it will go in the other direction.

What keeps me coming back to SRM is that the extremes for things like droughts, temperatures, flooding, hurricanes, sea ice – all the sorts of extremes that we see in the climate models – all those extremes go away when you apply geoengineering.

When we think about weather and climate, mostly we're worried about the extremes. You'll be worried if a hurricane

is going to flatten your house or if you're not going to get any
rain for five years... All of those things are ameliorated in our
climate models when you reduce the global warming.

James Haywood, Professor of Atmospheric Science,
University of Exeter

* * *

One last thought about SRM. The average global temperature
has been rising at about 0.2°C per decade, but it can bounce
around by as much as 0.25°C up or down from one year to the
next. The climate really is a 'chaotic system', and discerning the
real trend line is an exercise in hindsight that requires more
than a few years to become convincing. People imagine that
geoengineering would make a big splash as soon as you started
doing it, but the chaos in the system is so loud, the signal-to-
noise ratio so poor, that Francis Zwiers of the University of Victoria thinks in the first few years the geoengineers might find it
hard to prove their intervention is working.

Suppose you were able to fill the stratosphere with aerosols
at a reasonable cost. Suppose there was international agreement on that. And suppose you had the resources to do it for
five years. What would you know at the end of five years?

Would the climate change signal be detectable amidst
all of the noise in the climate system? And if you couldn't
demonstrate that it was effective, what would you do? Would
you proceed for another five years?

It's a really challenging problem. I don't think the problem
of detecting small signals in this very noisy system we live in
is being thought about very much, because the controversial
nature of the topic has made deployment seem a dramatic
gamble with immediate and massive consequences

The real mark of a successful deployment would be that nothing much changes at all. Immediate negative consequences would lead to a prompt shut-down, but clear evidence that the trend line of temperature had been shifted downwards could take a decade or more to become indisputable. After all, it took a century for the evidence of human influence on the climate to be confirmed at all.

In the context of geoengineering and also of big mitigation investments, the question is soon going to be: are these measures having their intended effect not just in reducing CO_2 concentrations in the atmosphere – that's easily measurable – but in terms of the resulting change in the weather? That's a much more challenging question, because it's a much harder signal to see.

Francis Zwiers

On reflection, however, this would be a good problem to have.

14

PLANETARY MAINTENANCE ENGINEERS

The larger the proportion of the Earth's biomass occupied by mankind and the animals and crops required to nourish us, the more involved we become in the transfer of solar and other energy throughout the entire system. As the transfer of power to our species proceeds, our responsibility for maintaining planetary homeostasis increases, whether we are conscious of the fact or not. Each time we significantly alter part of some natural process of regulation or introduce some new source of energy or information, we are increasing the probability that one of these changes will weaken the stability of the entire system... We shall have to tread carefully to avoid the cybernetic disasters of runaway positive feedback or of sustained oscillation...

This could happen if... man had encroached on Gaia's functional powers to such an extent that he had disabled her. He would then wake up one day to find that he had the permanent lifelong job of planetary maintenance engineer. Gaia would have retreated into the muds, and the ceaseless intricate task of keeping all the global cycles in balance would be ours. Then at last we should be riding that strange contraption, 'the spaceship Earth', and whatever tamed and domesticated biosphere remained would indeed be our 'life support system'... Assuming the present per capita use of energy, we can guess that at less than 10 billion people we should still be in a Gaian world. But somewhere beyond this figure, especially if the consumption of energy increases, lies the final choice of permanent enslavement on the prison hulk of the spaceship Earth, or gigadeath to enable the survivors to restore a Gaian world.

James Lovelock, *Gaia: A New Look at Life on Earth,* 1979

've referred a couple of times in passing to this bold prediction from James Lovelock's first book in 1979, but there it is
in full. I've known it almost by heart since I first read it, and
I intended to call this book *Planetary Maintenance Engineers*
until the publishers claimed that nobody would buy it with
that title and insisted on a change. You have to allow for the fact
that Lovelock was writing around forty-five years ago, when
birth rates were still high and economic growth rates in what
was then called the Third World were low, but the subsequent
very rapid growth in per capita energy use in Asia has delivered
us to exactly the future he foresaw by a slightly different route.
We are leaving the Gaian world of stable self-regulation behind.

It's not unmanageable yet. An average global temperature
around 1.2°C higher than pre-industrial is more than human
beings have seen at any point in the past 10,000 years, but not
so much higher that the world has become unrecognisable.
But the direction of travel is clear, the changes are accelerating
– and we human beings are gradually, reluctantly, half-unwit-
tingly slipping into the jobs as planetary maintenance engineers
that Lovelock predicted we would inherit. Those new jobs will
take up more and more of our time and resources, but if we are
both clever and lucky the crisis will only last a century or two.

Even ten years ago, few people realised it was going to be
like this. Most of those who were aware of climate change at all
assumed that the problem would be solved mainly by cutting
emissions, and that normal service could be restored once we
got that under control. Now our scientists are looking into
grinding up mountains and dumping them into the oceans to
combat acidification. They are learning to take carbon dioxide
out of the air and bury it underground. They are debating ways
to prevent our chemical wastes from destroying the ozone layer,
to disperse iron-salt aerosols from merchant ships in order to

remove methane from the atmosphere, to find microbes that will create artificial foods in bioreactors for animals and people so that we can return agricultural land to the wild. That's the sort of work that planetary maintenance engineers do: 'the ceaseless intricate task of keeping all the global cycles in balance'.

However, our alternatives are not as stark and inflexible as Lovelock feared: 'permanent enslavement on the prison hulk of the spaceship Earth or gigadeath to enable the survivors to restore a Gaian world'. It's not 1979 anymore, and many things have changed for the better. Half the technologies for dealing with the transition to a de-carbonised future that have been discussed in this book were not imagined four decades ago: cultivating algae as a food additive to inhibit methane production in ruminant livestock; devising ways to store electricity on a massive scale; stopping glaciers from sliding into the sea. No doubt some of them won't pan out, but our over-burdened planetary maintenance engineers will at least have a lot more tools to choose from.

The global political environment is better, too. Technology has helped a bit by endowing us with weapons so powerful that we dare not use them in our petty disputes. Those weapons might well have been used in the great ideological confrontation that divided the great powers into rival blocs for much of the twentieth century, but that conflict has now receded, and the wars of decolonisation that tormented the Global South in 1945–80 are long over. Backsliding is possible, terrorism is a constant nuisance, and the usual tribal/national disputes remain, but, compared to most of its previous history the world as a whole is a quite peaceful place. (I know the media constantly tell you the opposite, but that's their business model. They can't help it.)

Indeed, there is one particular regard in which the human

race is doing very well: since the Second World War, we have created genuinely global international institutions like the IPCC that go far beyond the old diplomacy that engaged individual countries or alliances over specific problems. These new organisations have a permanent legal existence, near-universal membership, professional staffs, and the job of regulating or at least supervising various activities at the global level. They also provide a precedent and a template for the institutions that we must build to cope with and, if possible, re-balance an increasingly disordered environment.

> The issue will become more and more about how we manage the fact that climate change is going to get worse and worse while we are doing more and more mitigation. No matter how fast we decrease carbon, the climate gets worse for another forty years. We've never tried to do politics like that before, and it will take some really far-sighted statesmen and women to realise that, if you're going to keep the deal going while everybody tightens their belts but no one sees returns, you're going to have to deal with the immediate consequences like food security, and show that there is some innate fairness.
>
> Climate change is unique. It raises enormous numbers of hard-security problems, and it has no hard-security solutions. In fact, the only solution it has is cooperation. Cooperation is difficult, it's emotionally draining, it's boring – just look at Europe – but it's the only way out. It's easy to be immature, to sit there and say 'we're going to defend our borders and be independent,' and blah, blah, blah. Much simpler politically, much simpler emotionally. It takes maturity to do cooperation. We need to realise that our hard-security future will only be preserved by doing this boring, drawn-out,

messy negotiation. We need to have a cadre or generation of leaders who get that. That's the hard yards. That's the World War Two of their generation. It's not glamorous, but it's incredibly important.

We spent hundreds of years learning how to handle the Industrial Revolution. We've got only one shot to handle climate change because, if we miss the target, the climate takes over itself, and all we've got left is adaptation, which will be brutal and ugly.

Nick Mabey, CEO, E3G (Third Generation
Environmentalism)

That interview took place a dozen years ago when I was writing *Climate Wars*, but Nick Mabey predicted the 2020s rather well. The emerging institution for global climate policy, the Conference of the Parties (COP), is developing its own culture and traditions, and from now on it will be an annual event. Its big meetings are typical human political gatherings where people know one another, reputation is important, and there is therefore real pressure to compromise. The generation born in the 1970s is in charge, they do get it, and so do those born in subsequent decades. Or at least enough of them do, and some of them are in the right places.

In terms of Adam Frank's three categories of emergent 'virtual civilisations' facing their inevitable climate crisis, this is moderately encouraging. We have collectively noticed that we have a problem, and we probably fall into one of the two categories that pick up on the issue early enough to do something about it. What remains unknown is how fast can we respond, and that is basically determined by culture at the species level. The question is not which local variety of human civilisation we belong to (they're all much alike, really, although we tend to see only the differences), but whether any species with the social and political heritage of the primate lineage can develop decision-making institutions that will produce a rapid and effective response to the climate challenge.

The jury will be out on that for quite a while yet, but the odds on our getting through this crisis without a human die-back or even a mass extinction are definitely better than they seemed in 1979. So it is worth our time to think our way at least through the years between now and 2035.

The Decade Ahead

By 2027 or 2028, the trend line will be unmistakable. Either we will be on course for 45 per cent cuts in annual greenhouse gas emissions by 2030, or we will be in trouble. Frankly, I'd take even 30 per cent emissions cuts by 2030 as a sign that we were finally getting onto the right course, but I don't believe in the Climate Fairy so I don't think that's likely either. A more plausible estimate is that the decade will end with something between 16 per cent more emissions than the 2010 level (the IPCC's 2021 'pessimistic' forecast) and, let's say, 16 per cent below it.[98] *Either* of those outcomes is failure, and we will all have some re-thinking to do.

We will probably have to do it amid mounting climate chaos. With the average global temperature rising by +0.27°C per decade, and with substantially unchanged emissions, it will almost certainly be back above +1.5°c by 2030. (I am assuming that peaks exceeding +1.5°C occurred during the El Niño episode that began in 2023, but that no irreversible tipping points were activated.) However, there will be another El Niño along in the late '20s or early 30s. It's cyclical: every three to seven years.

We won't yet be in completely unknown territory, but

the wildfires will be bigger and hotter, and the hurricanes/ typhoons/cyclones will be more powerful and destructive. Early 2023 saw the first Indian Ocean cyclone circle back and hit the east African coast for a second time, and by the end of the decade we may see the first Atlantic hurricanes re-curve all the way back across a warmer North Atlantic and hit Western Europe. As the temperature difference between the equator and the poles shrinks, the northern jet stream will slow further and meander even more widely, raising the prospect of hotter, longer heatwaves in summer, extreme cold spells in winter, and bigger, more frequent droughts and floods.

The global food supply may start to experience occasional large hiccups if atmospheric Rossby wave-5 or wave-7 patterns get stuck over the breadbasket regions of either or both hemispheres for longer periods of time, but real famines are unlikely except perhaps in the traditional venues. The Arctic sea ice will be melting more quickly, the glaciers will be sliding faster into the ocean, the sea level rise will speed up a bit, and every few years at least one low-lying, coastal big city will experience a New Orleans-style inundation. It could be a lot worse. Indeed, it *will* get a lot worse for a long time no matter what we do, because the linear feedbacks are always running and the warming process is cumulative.

By 2030, people will be sick and tired of the heat, the floods, the sheer unpredictability of daily life. The whole biosphere of conspiracy theorists from anti-vaxxers to QAnon adepts will still be with us, and they are likely to see geoengineering as an irresistibly attractive enemy. However, many more people will be better informed than today about climate issues and losing faith in the current mitigation-only policies.

Just as there are climate tipping points, there are also social tipping points, and 2030 may be when the argument that has been bubbling away in professional circles all through the previous decade comes to a boil and sucks in the whole world. It will be about solar geoengineering.

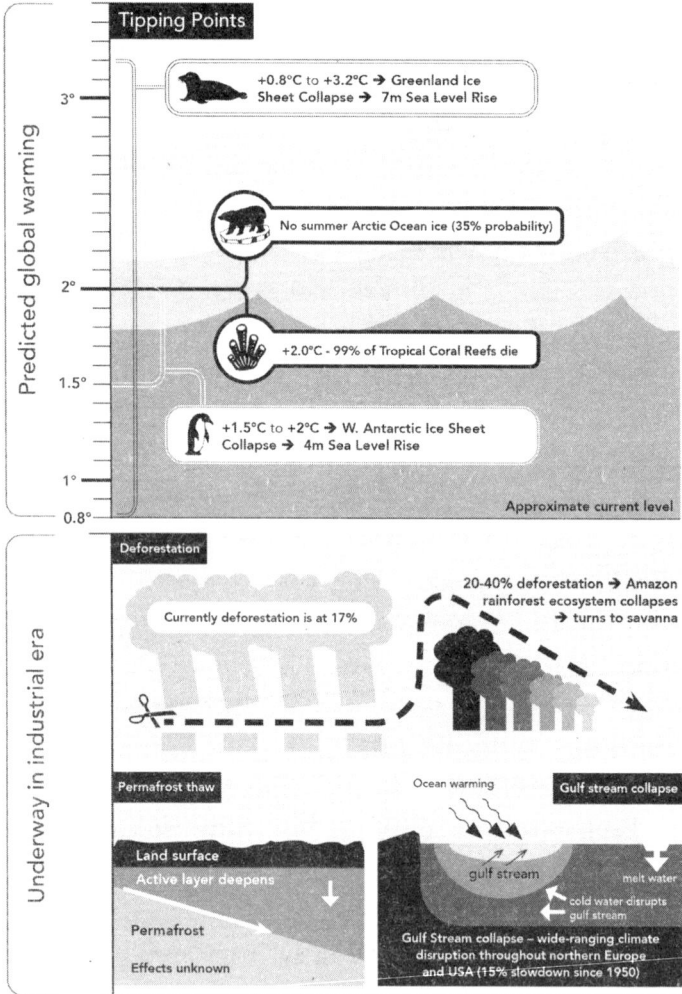

Tipping Points

+0.8°C to +3.2°C → Greenland Ice Sheet Collapse → 7m Sea Level Rise

No summer Arctic Ocean ice (35% probability)

+2.0°C - 99% of Tropical Coral Reefs die

+1.5°C to +2°C → W. Antarctic Ice Sheet Collapse → 4m Sea Level Rise

Approximate current level

Predicted global warming

3° — 2° — 1.5° — 1° — 0.8°

Deforestation

Currently deforestation is at 17%

20-40% deforestation → Amazon rainforest ecosystem collapses → turns to savanna

Permafrost thaw

Land surface
Active layer deepens
Permafrost
Effects unknown

Ocean warming

Gulf stream collapse

gulf stream
melt water
cold water disrupts gulf stream

Gulf Stream collapse – wide-ranging climate disruption throughout northern Europe and USA (15% slowdown since 1950)

Underway in industrial era

When discussing the known and unknown risks in solar geoengineering, we should devote a comparable effort to analysing the risks of *not* doing it. We are obliged by the circumstances to weigh a *balance* of risks pro and con. In reality, however, any decision to deploy solar geoengineering is more likely to be driven simply by evidence that tipping points are being crossed (which might make it too late, of course).

Probable 'linear' warming by 2035, based only on likely global emissions up to that date, ranges from +1.5°C to +1.8°C. These figures are 'fuzzy' and liable to amendment based on the latest research, particularly about southern mid-latitude cloud physics, but radical revisions are unlikely. The risk of crossing some tipping points will already exist from +1.5°C onwards, so a non-linear acceleration in the warming at this point is already possible.

The most difficult factor to calculate is the human one: when do food shortages, intolerable temperatures, monster storms and floods, waves of climate refugees, and the spread of failed states, pandemics and local wars cause so much damage and political instability that further global cooperation on climate issues collapses, or at least suffers a drastic decline? An average global temperature of +1.5°C seems low for such a catastrophe, but at +2.0°C or slightly higher, helped along by a cascade of tipping points, things could well start to fall apart.

> The clearest emergency would be if we were approaching a global cascade of tipping points that led to a new, less habitable, 'hothouse' climate state. Interactions could happen through ocean and atmospheric circulation or through feedbacks that increase greenhouse-gas levels and global temperature. Alternatively, strong cloud feedbacks could cause a global tipping point.
>
> We argue that cascading effects might be common. Research in 2018 analysed thirty types of regime shift

spanning physical climate and ecological systems, from collapse of the West Antarctic ice sheet to a switch from rainforest to savannah [in the Amazon and the Congo]. This indicated that exceeding tipping points in one system can increase the risk of crossing them in others. Such links were found for 45 per cent of possible interactions...

Now we are strongly forcing the system, with atmospheric CO_2 concentration and global temperature increasing at rates that are an order of magnitude higher than those during the most recent deglaciation. Atmospheric CO_2 is already at levels last seen around four million years ago, in the Pliocene epoch.

Timothy Lenton et al., 'Climate Tipping Points', *Nature*

Why It Has To Be Solar Engineering

The problem is too much heat and we know the solution: stop emitting greenhouse gases, then take the excess back out of the atmosphere to restore the atmosphere to what it was in, say, 1990. Unfortunately, that is a two-stage task: thirty years to stop the GHG emissions (Net Zero 2050), then another 50–100 years to remove the excess carbon dioxide and get back to an atmosphere and a temperature that suits us and all the other species that inhabit this planet.

You could spend a very long and unproductive day looking for a climate scientist who truly believes that we will stop the warming at +1.5°C. We might be able to stop at +2.0°C if everything goes just right and there are no surprises, but that's not the way things have gone in the past. We know that we risk crossing a number of tipping points on the way up to +2.0°C that could trigger a 'global tipping point': a far faster warming process that completely destabilises the biosphere. Unless we get our emissions down to net zero fast enough to keep the carbon dioxide in the atmosphere from rising past 450ppm (a fairly low probability), we will absolutely have to hold the heat down artificially while we get the necessary mitigation and negative emissions work done. The only way we have to do that is solar geoengineering. The risks involved in *not* doing it, at that point, will be much higher than the risks of doing it.

> If you don't want killer heat waves every year
> *Cool the planet*
> If you don't want waves of climate refugees
> *Cool the planet*
> If you don't the Amazon forest to collapse
> Cool the planet
> If you don't want >2 metres of sea level rise in
> your lifetime
> *Cool the planet*
> If you want to keep the coral reefs alive
> *Cool the planet*

It would have been much better to keep the planet at the right temperature by just cutting our greenhouse gas emissions, but we missed that exit several decades ago: around half of all human CO_2 emissions has been dumped into the air since 1990.

Unless further research reveals some hidden killer flaw, solar geoengineering is a safer course than letting the warming continue up to and even past +2.0°C. At present we're planning to fight this with one hand tied behind our back: we *know* the consequences of not doing it, and they are exceptionally grave.

Those who still believe that mitigation alone will do the trick should get on with it as fast as they can, but they should not try to block research on solar geoengineering (including open-air testing). The trend line of emissions and parts-per-million of

CO_2 will show who is right well before all we get to the deci-sion-point for deployment – probably around 2030 – but the solar geoengineering research needs to start accelerating soon so that we can be ready to deploy by then if necessary.

Marine Cloud Brightening (again)

Actually, if Marine Cloud Brightening proves to be cost-effec-tive, we should deploy it as soon as possible, even before 2030, because there's no lobby claiming that it poses a unique threat to climate stability. The doubt is only about whether the right size of reflective cloud droplets can be put into the clouds (or maybe even into cloudless air) to produce a worthwhile cooling effect. Spend on that research now, and we could be doing real MCB by 2030. A good place to start would be the Arctic, because the poles are where the warming is happening fastest, and in the past rapid warming there has had a big impact on overall climate stability.

There's no need to position the spray vessels right up in the high Arctic: cooling the air currents that transfer heat from the equator to the poles at any point on their journey would have about the same effect. Moreover, it would only need to be done in the few months each year when the sun is high enough to warm the Arctic. Spraying may also have to take place in the Southern Hemisphere in order to avoid moving the Inter-Trop-ical Convergence Zone around and shifting rainfall patterns.

As useful by-products of the Arctic cooling, glacial melt in Greenland would decrease, as would the slowing of the Gulf Stream – and the difference between temperatures in the Arctic and near the equator would start widening again, speeding up and therefore straightening out the Jet Stream – which might even nudge mid-latitude Northern Hemisphere weather back towards the traditional patterns. Or so one would hope.

Whether or not Marine Cloud Brightening can deliver these benefits, Stratospheric Aerosol Injection (SAI) will probably have to do the heavy lifting as soon as open-air experiments show that it is safe enough to go ahead. SAI is obviously a more fraught process, but the goal would be to inject enough aerosol each year to compensate for half of that year's warming. It would therefore be indirectly tied to the rate of progress in cutting emissions (lower emissions = less warming = less aerosol to counter half of that year's warming). It's no kind of cure, but by keeping the world cooler than it would otherwise be, it could keep societies intact and give them more time to deal with the real causes of the warming.

It could also provide us with an emergency tool to deal with nasty surprises. If a massive volcano erupts and causes a degree or two of cooling in one hemisphere for a couple of years, we would be able to stop injecting aerosol in that hemisphere and continue cooling the other one in order to maintain symmetry and keep the Inter-Tropical Convergence Zone where it belongs. If methane clathrates on the seabed should suddenly decompose and cause a sudden upward lurch in global temperature (as has happened during warming episodes in the past), we would at least have the ability to smooth it out a bit and limit the damage to people, animals and plants.

* * *

We are becoming planetary maintenance engineers far too soon, and we have to learn how to do it well in far too short a time. Nobody enjoys this prospect, but we are coming to accept the necessity. Hans Joachim Schellnhuber, the dean of German climate scientists and the founder and director emeritus of the Potsdam Institute for Climate Impact Studies, had always rejected climate engineering out of hand, but in 2021 I asked him about it again.

GD: You dismissed particularly Solar Radiation Management... If the alternative is as dire, as dreadful, as we're talking about, and there is no rapid-acting alternative, is your position still that this is dangerous nonsense?

HJS: In the end, if you really put me on the spot, and I would have to decide whether yes or no, and we have exhausted all other possibilities, and this would be a fair chance to avoid a hothouse Earth, if you manipulate the atmosphere for that then I would probably say grudgingly yes. Let's do it.

Yet still the nagging doubt remains: what can we do if we find that sulphur dioxide *does* expand the ozone holes? There's Marine Cloud Brightening, of course, but I'd very much like to have another SRM back-up as well – and it turns out that Robert Nelson may have one.

Atmospheric Salt

We started with the proposal in the 1990s by Edward Teller and his colleagues at Lawrence Livermore Lab, in which they said that you have nothing to fear from global warming because you could compensate for it by putting large quantities of aluminium oxide into the atmosphere. We had extensive experience with aluminium oxide and it's one of the hardest materials known. It's used for grinding glass, and it's not good for air-breathing organisms. So we said 'Out of the question. Could we find materials that have comparable reflective properties but would have a benign environmental impact?'

Robert Nelson

Bob Nelson worked at the renowned Jet Propulsion Lab at Caltech for forty years and wound up as Senior Research

Scientist. Much of his work focussed on how light reflects from planetary surfaces – 'Essentially we shine light on dirt and see how light scatters off it' – so when he retired he turned his attention to the SRM problem. He is exactly the right person to go looking for other, safer materials to do as good a job of reflecting incoming sunlight as aluminium oxide or sulphur dioxide, and he found several. The one he really likes is table salt. It has the right optical properties, and the question he's wrestling with now is whether salt can be manufactured in very small particle sizes, so that it floats in the air for a long time.

There was, however, one obvious problem with salt. It is sodium chloride, which in bright sunlight often 'photo-dissociates' into sodium and chlorine – and chlorine is precisely what we're worried may interact with CFCs to destroy ozone in the stratosphere. Rather than giving up, however, Nelson asked the less obvious question: why not put it into the air down here in the troposphere (the lower atmosphere below 20km), where all the living things are? It will still reflect the sunlight, it's cheap (no special stratospheric planes), and it isn't poisonous. We put it on our food.

> We had students calculate how much sodium chloride you would have to lift off the ground, distribute in the troposphere, and constantly replenish to counter one year's greenhouse gas emissions. It would be about 300 fully loaded 747 Super freighters every three months. If we're off by a factor of ten in either direction, it's still worth getting the engineers working on this.
>
> Robert Nelson

The idea of sulphur dioxide in the stratosphere took the lead because volcanoes have actually done that experiment for us: it cooled the planet, and it didn't hurt living things. However, if Bob Nelson's approach of working in the troposphere instead worked well, it would give us all the benefits of SO_2 in the stratosphere with none of the risks. Salt would require no new technology, it would be far cheaper, and it would be applied locally. If people in some particular locality don't want it there, just do it elsewhere.

> You can move the flights anywhere. You put a little extra here and you know it's going to have a consequence somewhere else on the surface, but you build it and shape it until you have the right configuration. If you try it off West Africa and rainfall decreases in the Amazon, then stop doing it off West Africa. It's somewhat like Marine Cloud Brightening, and it would require constant surveillance. We're taking control of a machine that is inherently willing to go unstable on us and we're trying to balance it.
>
> We're not saying it's a long-term solution, but it could provide temporary relief. It's analogous to what a physician does with a third-degree-burn victim. Give them morphine to stop the pain so they don't go into shock and die, then stabilise them and get their system back under control. It would be used on the assumption that society will control its emissions of greenhouse gases, get the environment under control. If it takes fifty years to turn that corner and get the temperature moving in the right direction, back down to what it was a hundred years ago, it would be worth doing the flights for fifty years.
>
> Robert Nelson

I'm including this idea not because I am sure it's the one key technology that will save us all from Hothouse Earth, but because it's an example of the original ideas for re-stabilising the environment that are now spilling out onto the table almost monthly. Some will work and some will not, but the sheer number of them is a reason to be hopeful.

Worldwide Fire Extinguisher?

What can we do in terms of emergency repair work if we find, in 2028 or 2033 or whenever, that we have unwittingly triggered a big climate feedback and the heating is rising at or beyond our worst-case projections? The likelihood of such an event is impossible to calculate, but those who know the history of past climate change would say that it's well above zero. Hysteresis is the hidden enemy: it's easy to change in one direction; much harder to come back in the opposite direction, so you need to nip these runaway feedbacks in the bud. We need something that is uncomplicated and easy to deploy quickly at scale. Cheap would also be nice, but is not indispensable.

Either Marine Cloud Brightening with Salter-style spray vessels or Nelson's tropospheric salt injections could work as a sort of worldwide fire extinguisher if it was approved for use and the machines were available in large numbers. But what if it isn't and they aren't? The one tried and tested emergency measure that we could turn on at a global scale within weeks is going back to burning high sulphate (3.5 per cent) fuel oil in those 60,000 commercial vessels that carry the world's trade. They wouldn't burn it in harbours or while steaming through densely populated coastal areas, of course – people's lungs matter, after all – but 90 per cent of their voyages are far from land. We could get back the ship tracks and one full degree of global cooling in a couple of months if we went all out on that; with more effort we could probably get 2 degrees of cooling or even more.

This is a capability that we should henceforward always have available at short notice, because we will be living in the danger zone permanently from now on. All we need to do is ensure that zealous accountants in the oil refineries don't dismantle the ability to produce large amounts of bunker oil on demand (I suspect that subsidies would help), and that all of those ships and all new ones retain the ability to switch to high-sulphur fuel if and when the world needs them to.

What Have Fossil Fuels Ever Done For Us?

The human race owes fossil fuels a great deal. Without them we would still be stuck on the treadmill of poverty, oppression and ignorance that was, until the eighteenth century, the lot of the majority of 'civilised' people at most times and in most places. It remains the lot of some people even today, of course, but a clear majority of the world's population now enjoys greater security, longer lives and regular meals. What made the difference was abundant, cheap energy.

For most of history, the only source of energy for most people was their own muscles and those of their domesticated animals. Wind began to be used to drive sailing ships in ancient Egypt more than 5,000 years ago, and the first reference to water-wheels for grinding grain is almost as old. The first windmills on land appeared about 1,300 years ago, but the familiar 'Dutch' type with vertical sails is only about 800 years old. And there was also gunpowder by then, but that was about it for machines that did useful work. All other work required muscle-power.

That goes a long way towards explaining why life in the past was so hard for most people: only the labour of many people could buy comfort for a few. Much of that labour was compulsory – feudalism or slavery – and there was almost invariably a hereditary ruling caste or class that lived off that labour. Right down to the eighteenth century, the skeletons of most men (and many women) show eroded joints due to a lifetime of heavy manual labour. Average lifespans were in the mid-thirties, most people were illiterate, and wars came and went as regularly, and as pointlessly, as weather systems. 'Progress' did happen from time to time, in a slow-motion sort of way (approximately eight centuries from the first saddle to the first stirrups), but, seen from the perspective of a single lifetime, the world was turbulent and yet unchanging: two steps forward, two steps back.

Coal liberated us from all that. Not right away – the steam-driven machines in early nineteenth-century England's 'dark Satanic mills' were harsher taskmasters than even the old landlords had been – and not everywhere at once, but gradually lifespans and living standards improved, which had never happened before. Oil joined coal in the mid-nineteenth century, and by the end of that century natural gas (methane) completed the Holy Trinity of fossil fuels that dominate the modern world.

It's doubtful that this 'modern' world of mass education, average lifespans into the seventies, and even exotic ideas such as democracy and feminism could ever have emerged from the long, brutal past without the assistance of fossil fuels. They carried a huge hidden cost, of course, but it was invisible to the early adopters and would not become apparent even to scientists until five or six generations later. At the time, it must have felt like the greatest good luck: just as firewood was becoming scarce and expensive because England's forests had mostly been cut down, they found millions of years' worth of fossilised forests underground that gave much better heat and seemed almost inexhaustible.

Human beings have been harnessing energy sources for 100,000 years. The discovery of fossil fuels was no different from that. We didn't trigger climate change because we're an evil virus on the planet, we triggered it as an inadvertent consequence of switching to a new energy source. Now that we know that it's leading to climate change, to do nothing about it would be totally our fault, but let's get out of the language of fault or blame.

If you look at it from an evolutionary, biospheric perspective, going extinct would be our own folly, nothing more, because if we cannot change our behaviour *that* would be why we don't make it through to the next step, if there is a next step. Assuming there are civilisations that have made it through this, we wouldn't be one of them. We would end up on the cosmic loser pile, rather than as one of the civilisations that got to go on and do interesting stuff. But we have to get away from this idea that human beings are somehow morally neglectful; that just our being here does a disservice to the planet. We are simply what the Earth is doing now.

I don't understand how people can't see this. A city is no different from a forest from the biosphere's point of view. We may like the cities for a particular reason, but from the biosphere's point of view it's just another evolutionary innovation, and the question is whether or not it gets to continue.

Adam Frank

We are where we are. Our two-century-long fossil fuel binge has put us in grave danger, but it has also delivered us to a place from which we can glimpse the possibility of a better future. Those two centuries allowed us to develop all the new energy technologies that would now allow us, if we choose, to dispense with fossil fuels and yet continue to enjoy the benefits of a high-energy civilisation. What we haven't yet developed is the values and behaviours that would enable us to cooperate beyond the local loyalties of nation, religion and ideology, and to give as much value to the rights and interests of future generations as we do to our own.

No fledgling civilisation emerges from its pre-civilised past already equipped with these values and behaviours. It develops them gradually, in the course of dealing with various crises that come with being a mass civilisation. Every national and cultural community on the planet has already gone through this process in order to create domestic peace and prosperity, so doing it once more at the global level is not an unimaginable task.

As a global society we clearly do not possess these qualities now, but to survive this emergency we will have to develop them. We won't embrace them quickly or happily – old habits die very hard – but when these seemingly impractical ideals become self-evident operational necessities, we may manage to choke them down in time to survive.

Final Words

Let's be honest. We want to preserve one particular climatic state among many potential ones, because it suits our needs and tastes. We don't want to head off into a 'hothouse' future, but neither do we want to return to the deep-frozen world of the Ice Ages, which is where we would be going around now if we hadn't unwittingly interfered in the climate long ago. We are players, not innocent bystanders, as are all the life forms on this planet, and we have inadvertently destabilised the climate we favour by two centuries of headlong industrialisation. Fully reversing the changes we made (if that is even possible) may require as much time and money as getting into this mess did.

Think of it as the final exam. (The mid-terms were about

developing nuclear weapons but not having an all-out nuclear war.) We have been building our civilisation for, in some places, 10,000 years, and it's now global in extent, but for most of that time we were really ignorant, semi-barbaric children. Only two centuries ago slavery was still a thriving institution, women were an inferior caste almost everywhere, resources were always scarce, and war was a normal way of doing business. Moreover, we understood almost nothing about how the planet we lived on actually worked.

Then we began burning fossil fuels, and for some people resources became abundant. Population and consumption soared, but so did science and knowledge. The nursery world that we thought we lived in – half playground, half battlefield, but unchanging and specially designed for human beings – turned out to be a fantasy. The real world was immensely old, it cared nothing for us, and there were many ways it could hurt us that we hadn't even imagined, from ice ages and asteroid strikes to runaway greenhouse warming and supernovas a hundred light-years away that could sterilise the planet. We were on our own, and it was time to grow up.

We haven't done so very badly. We have already managed to drop a lot of the baggage we were carrying from our long past, but some of it is really hard to get rid of: tribalism, for example, which normalises war but is also the social glue that holds most societies together. On the other hand, we have created new institutions that bind very large numbers of people together – democracy, universal education, mass media – and even some more-or-less functional international institutions. In terms of what we would need to pass the final exam and qualify for the long term, we are clearly acquiring some of the essential skills. The exam is not over. We're not doomed yet.

INTERVIEWEES

I must apologise to most of the people on this list for the subtitle of this book, which falsely conscripts them into the so far hypothetical profession of 'climate engineers'. I lost that argument, but I did want the book to come out and decided that this was the wrong hill to die on.

Jay Apt
Co-Director, Carnegie Mellon Electricity Industry Center, College of Engineering, Carnegie Mellon University, USA

Valentina Aquila
Professor, World Meteorological Organisation, American University, USA

Govindasamy Bala
Professor, Centre for Atmospheric and Oceanic Sciences, Indian Institute of Science, Bangalore, India

Ulrike Burkhardt
Researcher, Institute of Atmospheric Physics, DLR, German Aerospace Centre, Germany

Ken Caldeira
Senior Scientist (emeritus), Department of Global Ecology, Carnegie Institution for Science, USA

Donald Canfield
Professor of Ecology, Institute of Biology, University of Southern Denmark, Denmark

Galina Churkina
Professor, Technische Universität Berlin, Germany

Emily Cox
Research Associate, School of Psychology, Cardiff, Wales

Peter Cox
Department of Mathematics, College of Engineering, Mathematics and Physical Sciences, University of Exeter, UK

Felix Creutzig
Head of group, Land Use, Infrastructure and Transport, Mercator Research Institute on Global Commons and Climate Change (MCC), Berlin, Germany

Greg Dipple
Professor of Geology, University of British Columbia, Canada

Scott Doney
Joe D. and Helen J. Kington Professor in Environmental Change, Dept. of Environmental Sciences, University of Virginia, USA

Tamsin Edwards
Reader in Climate Change, King's College London, UK

John Fasullo
Project Scientist III, Climate Analysis, National Center for Atmospheric Research NCAR, USA

Peter Fraenkel
CEO, Gravitricity Ltd, UK

Jennifer A. Francis
Senior Scientist, Woodwell Climate Research Center, Falmouth, Massachusetts, USA

Adam Frank
Professor, Department of Physics and Astronomy, Univ. of Rochester NY, USA

Alan Gadian
Professor, National Centre for Atmospheric Sciences, Leeds University, UK

Dieter Gerten
Research Group Leader for Earth Modelling, Potsdam Institute for Climate Impact Research, Germany

James E Hansen
Director, Program on Climate Science, Awareness and Solutions of the Earth Institute at Columbia University, USA

Cheryl S. Harrison
Assistant Professor, Department of Oceanography and Coastal Sciences, Louisiana State University, USA

Stuart Haszeldine
Professor of Carbon Capture and Storage, School of Geosciences, University of Edinburgh, UK

James Haywood
Professor of Atmospheric Science, College of Engineering, Mathematics and Physical Sciences, University of Exeter, UK

Hermann J. Heipieper
Senior Scientist, Helmholtz Centre for Environmental Research, Leipzig, Germany

John Holdren
Teresa and John Heinz Professor of Environmental Policy at Harvard's Kennedy School of Government, CoDirector of the School's Science, Technology, and Public Policy program, USA

Matthias Honegger
Research Associate, Climate Engineering in Science, Society and Politics, Institute for Advanced Sustainability Studies, Potsdam, Germany

Hugh Hunt
Reader in Engineering Dynamics & Vibration, Mechanics, Materials and Design, Cambridge University, UK

Peter Irvine
Lecturer in Climate Change & Solar Geoengineering, University College London, UK

Rajeev Jaiman
Associate Professor and NSERC/Seaspan Industrial Research Chair, Department of Mechanical Engineering, University of British Columbia, Canada

Irene Thor Jeremiassen
Ilulissat, Greenland

Bowie Keefer
Co-founder and Senior Vice President Business Development at Svante Inc. Burnaby, British Columbia, Canada

David Keith
Gordon McKay Professor of Applied Physics, Paulson School of Engineering and Applied Sciences and Professor of Public Policy, John F. Kennedy School of Government, Harvard University, USA *and* Chief Scientist, Carbon Engineering Squamish, British Columbia, Canada

David P. Keller
Senior scientist in the Research Unit Biogeochemical Modelling, GEOMAR Helmholtz Centre for Ocean Research, Germany

Luke Kemp
Research Associate, Centre for the Study of Existential Risk, Cambridge, UK

Anton Keskinen and Ada Koistinen
Head of Strategy at Operaatio Arktis, Climate Activists, Helsinki, Finland

Joakim Kjellsson
Researcher, GEOMAR Helmholtz Centre for Ocean Research, Division of Maritime Meteorology, Germany

Roel van Klink
Researcher, German Centre for Integrative Biodiversity Research, Leipzig, Germany

Kai Kornhuber
Researcher, Earth Institute, Hogan Hall MC 3277, Columbia University, USA

Ben Kravitz
Assistant Professor, Department of Earth and Atmospheric Sciences, College of Arts and Sciences, Indiana University Bloomington;
Earth System Modeling, Pacific Northwest National Laboratory, Richland, Washington State, USA

Lee Kump
Dean, College of Earth and Mineral Sciences and Professor of Geosciences, Pennsylvania State University, USA

William Lamb
Researcher, Mercator Research Institute on Global Commons and Climate Change (MCC), EUREF Campus 19, Berlin, Germany

Robin Lamboll
Research Fellow, Faculty of Natural Sciences, Centre for Environmental Policy, Imperial College London, UK

Mark Lawrence
Scientific Director, Institute for Advanced Sustainability Studies, Potsdam, Germany

Tim Lenton
Director Global Systems Institute, University of Exeter, UK

Johan Lilliestam
Research Group Leader, Institute for Advanced Sustainability Studies, Potsdam, Germany

James Lovelock
Dorset, UK

Daniel Lunt
Professor of Climate Science, School of Geographical Sciences, University of Bristol, UK

Douglas MacMartin
Senior Research Associate, Sibley School of Mechanical and Aerospace Engineering, Cornell University, USA

Jochem Marotzke
Director of the department of climate variability, Max Planck Institute for Meteorology, Hamburg, Germany

Nils Matzner
Research Fellow, Munich Centre for Technology in Society, Germany

Justin McClellan
Business Development and Strategy Lead – Aircraft Electrification, BAE Systems, Nashua, New Hampshire, USA

Ilona Mettianan
Researcher, Arctic Centre of the University of Lapland, Finland

Manfred Milinski
Emeritus Scientific Member, Max Planck Institute of Evolutionary Biology, Germany

John C. Moore
Chief Scientist, College of Global Change and Earth System Science, Beijing Normal University, China

Robert Nelson
Research Scientist, Planetary Science Institute, Pasadena, California, USA

Richard J. Nevle
Deputy Director, Earth Systems Program, Yang and Yamazaki Environment and Energy Building, Stanford University, USA

Simon Nicholson
Associate Professor, School of International Service, American University, Washington, USA

Ulrike Niemeier
Scientist, Max Planck Institute for Meteorology, Hamburg, Germany

Euan Nisbet
Emeritus Professor (Earth Sciences), Department of Earth Sciences, Royal Holloway, University of London, UK

Ilissa Ocko
Senior Climate Scientist, Barbra Streisand Chair of Environmental Studies, Environmental Defense Fund, USA

Romaric C. Odoulami
AXA Research Chair, African Climate Risk, University of Cape Town, South Africa

Steve Oldham
Managing Director, Carbon Engineering, Squamish, British Columbia, Canada

Andreas Oschlies
Head of Research Unit Biogeochemical Modelling, GEOMAR Helmholtz Centre for Ocean Research, Germany

Jyoti Parikh
Executive Director, Integrated Research and Action for Development (IRADe), New Delhi, India

Andrew Parker
Senior Research Fellow, School of Earth Sciences, Bristol, UK

Hosea Olayiwola Patrick
Postdoctoral Fellow, School of Built Environment and Development Studies, Earth System Governance Project, University of KwaZulu-Natal, South Africa

Juha-Pekka Pitkänen
CEO, Solar Foods Oy, Finland

Julia Pongratz
Chair of Physical Geography and Land-Use Systems, Ludwig-Maximilians-Universität, Munich, Germany

Stefan Rahmstorf

Co-Chair, Earth System Analysis, Potsdam Institute for Climate Impact Research, Germany

David Reichwein

LLB Fellow, Research & Policy Analysis, Legal Team, Ecologic Institute, Berlin, Germany

Katherine Richardson

Principal Investigator, Centre for Macroecology, Evolution, and Climate, University of Copenhagen, Denmark

Alan Robock

Distinguished Professor, Department of Environmental Sciences, School of Environmental and Biological Sciences, Rutgers University, New Jersey, USA

Johan Rockström

Director, Potsdam Institute for Climate Impact Research (PIK), Germany; Professor, Stockholm Resilience Centre, Sweden

Joeri Rogelj

Lecturer in Climate Change and the Environment, Faculty of Natural Sciences, The Grantham Institute for Climate Change, Imperial College London, UK

Karen Rosenlof

Senior Scientist for Climate and Climate Change, NOAA Chemical Sciences Laboratory, Boulder, Colorado, USA

William Ruddiman

Professor Emeritus, Department of Environmental Sciences, University of Virginia, USA

Stephen Salter

Emeritus Professor of Engineering Design, School of Engineering, University of Edinburgh, UK

Stefan Schäfer

Research Group Leader, Climate Engineering in Science, Society and Politics, Institute for Advanced Sustainability Studies Potsdam, Germany

Hans Joachim Schellnhuber CBE

Director Emeritus, Potsdam Institute for Climate Impact Research (PIK), Germany

Gavin A. Schmidt
Director, NASA Goddard Institute for Space Studies, New York, NY, USA

Christina Schoof
Professor, Earth, Ocean and Atmospheric Sciences, University of British Columbia, Canada

Troels Schönfeldt
CEO, Seaborg Corporation, Copenhagen, Denmark

Wake Smith
Senior Fellow, Mossavar-Rahmani Center for Business and Government, Harvard Kennedy School, USA

William Steffen
Senior research fellow, Stockholm Resilience Centre, Stockholm University, Sweden; Emeritus Professor, Fenner School of Environment and Society, Australian National University, Canberra ACT, Australia

Jan Steckel
Mercator Research Institute on Global Commons and Climate Change (MCC) GmbH, Berlin, Germany

Rowan Sutton
Director of Climate Research, National Centre for Atmospheric Science, Department of Meteorology, Reading, UK

Abu Syed
Senior Fellow, Centre for Advanced Studies, Dhaka, Bangladesh

Jau Tang
Professor, Institute of Technological Sciences, Department of Engineering, Wuhan University, China

David Thornalley
Associate Professor, Department of Geography, University College London, UK

Simone Tilmes
Project Scientist II, Atmospheric Chemistry Observations, and Modeling Laboratory, National Center for Atmospheric Research, Boulder, CO, USA

Sandro Vattioni
Professor, Institute of Atmospheric and Climate Science, ETH, Zurich, Switzerland

Steve Vavrus
Senior Scientist, Nelson Institute Center for Climatic Research, University of Wisconsin, USA

Mikkel Vejerønsbo
Head of Communications, Seaborg Corporation, Copenhagen, Denmark

Martin Voigt
Post Doctoral Researcher, Carbfix Technology, University of Iceland, Iceland

Bob Ward
Policy and Communications Director, Grantham Research Institute on Climate Change and Environment, London School of Economics, UK

Matthew Watson
Reader in Natural Hazards, School of Earth Sciences, Bristol University, UK

Jessica Whiteside
Associate Professor, Ocean and Earth Science, Southampton, UK

Phil Williamson
NERC Research Coordinator, Hubert Lamb Building, School of Environmental Sciences, University of East Anglia, UK

Michael Wolovick
Postdoctoral Research Scientist, Alfred Wegener Institute, Helmholz, Germany

Ole Wroldsen
Civil Engineer, Marine Scientist, Orca Solutions, Norway

Francis Zwiers
Senior Research Scientist and Chief of the Canadian Centre for Climate Modelling and Analysis, University of Victoria, Canada

ENDNOTES

1 See, for example, European Academies Science Advisory Council, 'Negative emission technologies: What role in meeting Paris Agreement targets?', EASAC Policy Report 35 (February 2018).

2 The entire paper is available free on www.pnas.org/cgi/doi/10.1073/pnas.1810141115

3 For a critical review of this article, see Richard Betts, 'Hothouse Earth: here's what the science actually does – and doesn't – say', *The Conversation*, August 9, 2018, https://theconversation.com/hothouse-earth-heres-what-the-science-actually-does-and-doesnt-say-101341

4 Australia, Belgium, Denmark, Germany, Netherlands, Sweden, United Kingdom and United States.

5 Altmetric, or altmetric.com, is a data science company that tracks where published research is mentioned online, and provides tools and services to institutions, publishers, researchers, funders and other organisations to monitor this activity, commonly referred to as altmetrics.

6 The 2009 article: Johan Rockström, Will Steffen,Timothy M. Lenton, Marten Scheffer, Carl Folke, Hans Joachim Schellnhuber, James Hansen, Katherine Richardson, Paul Crutzen Jonathan Foley et al. 'Planetary Boundaries: Exploring the Safe Operating Space for Humanity', Ecology and Society, Vol. 14, No. 2 (Dec 2009) https://www.jstor.org/stable/26268316

The 2015 article: Will Steffen,* Katherine Richardson, Johan Rockström, Sarah E. Cornell, Ingo Fetzer, Wim de Vries, Cynthia A. de Wit, Carl Folke, Dieter Gerten, Veerabhadran Ramanathan, Belinda Reyers, et al., 'Planetary boundaries: Guiding human development on a changing planet', *Science*, 15 Jan 2015, Vol. 347, Issue 6223. DOI: 10.1126/science.1259855

The 2023 article: Katherine Richardson, Will Steffen, Jonathan F. Donges, Govindasamy Bala et al., 'Earth beyond six of nine planetary boundaries', *Science Advances*, 13 Sep 2023, Vol 9, Issue 37 DOI: 10.1126/sciadv.adh2458

7 This name, although widely used, is not yet officially recognised by the guardians of the stratigraphic flame.

8 *Trajectories*, pp. 8253–54

9 Yeon-Hee Kim, Seung-Ki Min, Nathan P. Gillett, Dirk Notz & Elizaveta Malinina, 'Observationally-constrained projections of an ice-free Arctic even under a low emission scenario', *Nature Communications* vol. 14, No. 3139 (2023).

10 United Nations Development Programme, *The People's Climate Vote*, 26 January 2021

11 Robin D. Lamboll, Joeri Rogelj et al., 'Assessing the size and uncertainty of remaining carbon budgets', *Nature Climate Change* vol. 13, pp1360–1367 (2023) https://doi.org/10.1038/s41558-023-01848-5

12 Gavin A Schmidt and Adam Frank, 'The Silurian hypothesis: would it be possible to detect an industrial civilization in the geological record?', *International Journal of Astrobiology*, Vol. 18 , No 2 , April 2019 , pp. 142 - 150.DOI: https://doi.org/10.1017/S1473550418000095

13 Estimates of the current number of species range from five to fourteen million. G. Miller; Scott Spoolman (2012). *Environmental Science – Biodiversity Is a Crucial Part of the Earth's Natural Capital*. Cengage Learning, p. 62. But individual species come and go quite quickly, in terms of geological time: over 99% of the

estimated five billion different species of life forms that ever lived on Earth are extinct. Stearns, Beverly Peterson; Stearns, Stephen C., *Watching, from the Edge of Extinction*. Yale University Press, 2000

14 The oceans remained largely anoxic despite the celebrated Great Oxygenation Event of about 2–2.4 billion years ago, which only raised the level of free oxygen in the atmosphere from essentially zero to ~2%. It was the evolution of modern plants that started raising the level of oxygen in the oceans and the atmosphere around 800 million years ago. By the time of the Cambrian Explosion (541 mya), when the oxygen may have reached about the modern level (21%) and a multitude of animal species appeared, anaerobic bacteria had already been forced to retreat into the muds.

15 Lee R. Kump, Alexander Pavlov and Michael A. Arthur, 'Massive Release of Hydrogen Sulfide to the Surface Ocean and Atmosphere During Intervals of Oceanic Anoxia', *Geology*, May 2005, vol. 33, no. 5, pp. 397-400; http://geology. geoscienceworld.org/cgi/content/abstract/33/5/397

16 An alternative number for the cumulative carbon emitted during the second phase of the end-Permian event, which coincides with the main extinction, is 21 trillion tonnes of carbon – 30 times more than the cumulative anthropogenic carbon emitted between 1750 and 2020 (690 billion metric tons of carbon). Yuyang Wu, Ying Cui et al., 'Volcanic CO_2 degassing postdates carbon emission during the end-Permian mass extinction', *Science Advances*, 2023 Feb 15;9(7):eabq4082. DOI: 10.1126/sciadv.abq4082

17 Information and quotes in this chapter, unless otherwise indicated, is from the paper 'Global Warming' (James Hansen et. al., 'Global Warming in the Pipeline', Oxford Open Climate Change, Vol 3 No 1, 2023,https://doi.org/10.1093/oxfclm/kgad008), or from an interview by GD on 20 November 2023.

18 Video moderated by Jeffrey Sachs: 'An Intimate Conversation with Leading Climate Scientists to discuss Groundbreaking New Research on Global Warming', Nov 2, 2023. https://www.youtube.com/watch?v=NXDWpBlPCY8

19 Alan M. Seltzer et al., 'Widespread six°Celsius cooling on land during the Last Glacial Maximum', Nature, 593, pages 228–232 (2021); Robert Monroe, 'Climate Colder On Land During Last Ice Age Than Thought', UC San Diego/Scripps Institution of Oceanography, 12 May 2021, https://scripps.ucsd.edu/news/climate-colder-land-during-last-ice-age-thought;

20 'IMO2020 fuel oil sulphur limit - cleaner air, healthier planet', International Maritime Organisation, 28 January 2021; James Dineen, 'Cleaner shipping emissions may have warmed the planet – but only a bit', New Scientist, 10 August 2023.

21 Jake Spring and David Stanway, 'Climate's Catch-22: Cutting pollution heats up the planet', *Reuters*, 2 November 2023.

22 Video moderated by Jeffrey Sachs…

23 Michael E. Mann, 'Comments in New Article by James Hansen', https://michaelmann.net/content/comments-new-article-james-hansen

24 Hugh Hunt and Robert Chris, 'The disagreement between two climate scientists that will decide our future', *The Conversation*, 8 December 2023 https://theconversation.com/the-disagreement-between-two-climate-scientists-that-will-decide-our-future-21775925 https://www.iea.org/energy-system/renewables/solar-pv#overview

26 *ibid*

27 Sophie Mellor, 'The U.K. went all in on wind power. Here's what happens when it stops blowing', *Fortune*, 16 September 2021

28 Jovana Radulovich et al., 'Pumped Thermal Energy Storage Technology (PTES): Review', *Thermo* 2023, 3(3), 396-411 https://doi.org/10.3390/thermo3030024; Environmental and Energy Study Institute, *Fact Sheet | Energy Storage*, February 22, 2019.

29 'In a First, Caltech's Space Solar Power Demonstrator Wirelessly Transmits Power in Space', *Caltech*, June 01, 2023; 'Space Based Solar Power: De-risking the pathway to Net Zero', Frazer-Nash Consultancy Ltd. for UK Department of Business, Energy, and Industrial Strategy, September 2021; Jonathan Leake, 'Tim Peake backs plans for solar farms in space', *The Telegraph*, 17 Sep 2023.

30 Michael Shellenberger, 'Why Renewables Advocates Protect Fossil Fuel Interests, Not The Climate', *Forbes*, 28 March 2019

31 US Energy Information Administration, 'Today in Energy: EIA projections show hydro growth limited by economics not resources', 10 July 2014.

32 'Geothermal energy has potential to fulfil much of UK's current heating need', *UK Parliament Environmental Audit Committee*, 20 October 2022.

33 *TIME* Magazine 7 August 2023; info@fervoenergy.com

34 Robert W. Howarth and Mark Z. Jacobson, 'How green is blue hydrogen?', *Energy Science and Engineering*, Volume 9, Issue10, October 2021, pp. 1676-1687 https://doi.org/10.1002/ese3.956

35 International Energy Agency, *Global Hydrogen Review* 2023 (September 2023). Three-quarters of these projects are 'green hydrogen' (electrolysis with low-emission electricity); the remaining quarter are 'blue hydrogen'. More than half of this planned production is still in the early stages of development; about half of all current green hydrogen production is in China.

36 Jillian Ambrose, 'Oil firms made 'false claims' on blue hydrogen costs, says ex-lobby boss', *Guardian*, 20 August 2021.

37 Eric Hand, 'Hidden Hydrogen', *Science*, Vol 379, Issue 6633. doi: 10.1126/science.adh1460

38 'Green Steel Tracker', https://www.industrytransition.org/green-steel-tracker/

39 All dates and details from the United States Environmental Protection Agency rules, but comparable measures have been taken by other signatories on similar timetables.

40 Minnesota Pollution Control Agency, https://www.pca.state.mn.us/air/chlorofluorocarbons-cfcs-and-hydrofluorocarbons-hfcs

41 Georg Bieker, 'A Global Comparison of the Life-cycle Greenhouse Gas Emissions of Combustion Engine and Electric Passenger Cars', *The International Council on Clean Transportation*, 20 July 2021

42 Josh Gabbatiss, 'Explosive growth means one in three new cars will be electric by 2030, IEA says', *Carbon Brief*, 26 April 2023

43 Stefanie Schumann, 'How "Green" is Very Low Sulphur Fuel Oil (VLSFO)?', https://help.fleetmon.com/en/articles/4521749-how-green-is-very-low-sulphur-fuel-oil-vlsfo; Gavin Maguire, 'The ironic side effects of the rapid global energy transition', *Reuters*, 10 March 2023.

44 Susan van Dyk, 'Direct air capture CO2 for aviation e-fuels production faces many obstacles finds E4Tech study', https://www.greenairnews.com/?p=1334, 13

July 2021; https://bc.ctvnews.ca/b-c-facility-aims-to-make-vehicle-fuel-from-carbon-pulled-out-of-the-atmosphere-1.5624124

45 Jonathan Welsh, 'Sustainable Aviation Fuel Production Tripled in 2022: IATA', *Flying*, 3 January 2023.

46 Jane O'Malley, Nikita Pavlenko, Stephanie Searle, 'Estimating sustainable aviation fuel feedstock availability to meet growing European Union demand', *ICCT Working Paper 2021-13*, March 2021

47 https://e360.yale.edu/features/how-airplane-contrails-are-helping-make-the-planet-warmer https://doi.org/10.1038/d41586-021-02990

48 Fusion Industry Association, 'A Bold Decadal Vision for Fusion Energy: FIA Participates in White House Fusion Summit', 17 March 2022. https://www.fusion-industryassociation.org/a-bold-decadal-vision-for-fusion-energy-fia-participates-in-white-house-fusion-summit/

49 Jeff Tollefson, 'Top climate scientists are sceptical that nations will rein in global warming', *Nature* 599, 22-24 (2021) https://doi.org/10.1038/d41586-021-02990

50 *Guardian*, 22 May 2018.

51 Lyrics reproduced by permission of the Flanders & Swann Estates.

52 Emissions data in this and the following three paragraphs are all taken from 'A Global Breakdown of Greenhouse Gas Emissions by Sector' in *OurWorldinData. org*, 2020 www.visualcapitalist.com/a-global-breakdown-of-greenhouse-gas-emissions-by-sector/

53 Beef (26.61kg. of CO_2) and lamb (25.58kg.) are off the scale because these are ruminants that emit methane, whereas pigs (5.77kg), poultry (3.65kg.) and fish (3.49kg.) do not. 'Here's the real impact of the food we eat on the environment', *World Economic Forum*, Dec. 2016, https://www.weforum.org/agenda/2016/12/your-kitchen-and-the-planet-the-impact-of-our-food-on-the-environment

54 5% of global greenhouse gas emissions comes from animals burping and farting methane (and 0.8% from their manure, but leave that out.) Add a little bit of the right inhibitor to the animals' feed, and you reduce those emissions by up to four-fifths (30% to 82%). Net share of global emissions from this source falls to 1%. In theory, 4% saved. In practice much less, but still...

55 Ben Elgin, 'Why Won't Companies Use This Quick Fix to Reduce Cow Methane Emissions?', *Bloomberg*, 28 June 2023. https://www.bloomberg.com/news/features/2023-06-28/this-quick-fix-reduces-methane-emissions-from-cow-burps?leadSource=uverify%20wall

56 G. Supran, S. Rahmstorff, N. Oreskes, 'Assessing ExxonMobil's global warming projections', *Science*, Vol. 379, No. 6628, 13 Jan 2023. DOI: 10.1126/science.abk0063

57 Statistics from Kate Lyons, 'Cutting food waste by 25% could feed world', *Guardian*, 12 August 2015.

58 For further discussion of the various methane removal methods being considered, see https://www.sparkclimate.org/methane-removal-approaches; and Tingzhen Ming, Wei Li, Qingchun Yuan, Philip Davies et al., 'Perspectives on removal of atmospheric methane', *Advances in Applied Energy*, Vol. 5, February 2022. doi.org/10.1016/j.adapen.2022.100085

59 The 'Carrington Event' was the most intense geomagnetic storm in recorded history, in September 1859, just a few months before the solar maximum of 1860. The coronal mass ejection or solar flare happened to intersect just the right part

of Earth's orbit as we traveled around the sun, causing intense auroral displays even quite close to the equator and shutting down the world's then-new telegraph systems. A similar event today could cause an 'internet apocalypse'. (The next solar maximum is 2025.)

60 Adam Baylin-Stern and Niels Berghout, "Is carbon capture too expensive?", *International Energy Agency*, Commentary — 17 February 2021

61 The amount of CO_2 being captured today is 43 million tons, or 0.1% of global emissions. If all the likely projects that have been announced come online, there would be 279 million tons of CO_2 (0.6% of today's emissions) captured every year by 2030, mostly for the power sector, for the manufacture of low-carbon hydrogen and ammonia, or to abate emissions from industrial sources. Hardly an impressive amount, given that the steel and cement industry along account for 13% of global emissions. Bloomberg NEF, 'Global Carbon Capture Capacity Due to Rise Sixfold by 2030', 18 October 2022. https://about.bnef.com/blog/global-car-bon-capture-capacity-due-to-rise-sixfold-by-2030/#:~:text=By%202030%2C%20 most%20capture%20capacity,or%200.1%25%20of%20global%20emissions. Note that this is not 'Direct Air Capture', a fledgling technology that is just starting to build its first one-million-tonnes-a-year plants in the United States. Neither is it in-tended to 'abate' CO_2 emissions from ordinary manufacturing that use fossil fuels, although attempts to hijack it for those purposes are underway.

62 'Do One Better!' podcast No. 83 with Alberto Lidji, 10 April 2020.

63 Oliver Milman, 'The world's biggest carbon capture facility is being built in Texas. Will it work?', *Guardian*, 12 September 2023

64 This is a new technology, but potentially very important, as it is extremely simple, easily scalable, and addresses both the CO_2 and the ocean acidification problems. In 2024 Captura will start building a pilot plant in eastern Quebec, in partnership with Montreal-based carbon removal and storage venture Deep Sky, that takes advantage of the province's abundant hydropower. The ultimate goal is a million-tonne-a-year plant.

65 Danielle Radin, 'UCLA engineers develop solution to reduce atmospheric carbon dioxide', *KCAL News,* 12 April 2023

66 Jennifer L., 'Bottom Trawling Fishing Emits as Much Carbon Emission as Avi-ation', *Carbon Credits*, 9 June 2022. A later paper has suggested that the emissions are not so large (https://www.hw.ac.uk/news/articles/2023/lyell-centre-scientists-cast-doubt-on.htm)

67 National Ocean and Atmospheric Administration, U.S. Dept. of Commerce, 'Ocean Acidification'.

68 Heike K. Lotze, 33 other authors and Boris Worm, 'Global ensemble projec-tions reveal trophic amplification of ocean biomass declines with climate change', *PNAS*, June 11, 2019, 116 (26) https://doi.org/10.1073/pnas.1900194116

69 Daniel Pauly and Dirk Zeller, 'Catch reconstructions reveal that global marine fisheries catches are higher than reported and declining', *Nature Communications*, 7, Article number: 10244 (2016) https://www.nature.com/articles/ncomms10244

70 Villy Christensen, Marta Coll, Chiara Piroddi, Jeroen Steenbeek, Joe Buszows-ki, Daniel Pauly, 'A century of fish biomass decline in the ocean', Contribution to the Theme Section 'Trophodynamics in marine ecology', *Marine Ecology Progress Series*, MEPS 512:155-166 (2014) - DOI: https://doi.org/10.3354/meps10946

71 Rosamond L. Naylor, Jane Lubchenco, Sandra E. Shumway et al., 'A 20-year retrospective review of global aquaculture', *Nature*, 591, 551–563 (2021). https://

doi.org/10.1038/s41586-021-03308-6

72 Naylor et al., *ibid.*, and Björn Kok, Wesley Malcorps, Michael F.Tlusty, Mahmoud M. Eltholth, Neil A. Auchterlonie, David C. Little, Robert Harmsen, Richard W. Newton, and Simon J.Davies, 'Fish as feed: Using economic allocation to quantify the Fish In : Fish Out ratio of major fed aquaculture species,' *Aquaculture*, Vol. 528, 15 November 2020, 735474

73 'Global and European sea level rise', *European Environment Agency*, 18 Nov 2021 https://www.eea.europa.eu/ims/global-and-european-sea-level-rise

74 Sjoerd Groeskamp and Joakim Kjellsson, 'Northern Europe Enclosed: Engineering a Solution to Sea Level Rise', *Bulletin of the American Meteorological Society*, Vol. 101: No. 11, pp. 971–974, 1 Nov 2020. DOI: https://doi.org/10.1175/BAMS-D-19-0145.A

75 Andrew Lockley, Michael Wolovick, Bowie Keefer, Rupert Gladstone, Li-Yun Zhao and John Moore, 'Glacier geoengineering to address sea-level rise: A geotechnical approach' *Advances in Climate Change Research* 11(133) December 2020 DOI:10.1016

76 Paul Voosen, 'Ice shelf holding back keystone Antarctic glacier within years of failure', *Science,* 13 Dec 2021, Vol. 374, No 6574, pp. 1420-1421 DOI: 10.1126/science.acz9821

77 Leon Fuerth is a former diplomat and professor of international relations at George Washington University. He served as National Security Adviser to Vice-President Gore in 1993-2001. The paper was published by the Center for Strategic and International Studies in November 2007, and unfortunately there's almost nothing in it that needs to be changed today.

78 Colin Raymond, Tom Matthews, and Radley M. Horton, 'The emergence of heat and humidity too severe for human tolerance', *Science Advances* 8 May 2020: Vol. 6, no. 19, eaaw1838 DOI: 10.1126/sciadv.aaw1838

79 Population Division of the United Nations Department of Economic and Social Affairs (UNDESA), quoted in *World Resources Institute* paper by Tim Searchinger, Craig Hanson, Richard Waite, Brian Lipinski, George Leeson and Sarah Harpe, 'Achieving Replacement Level Fertility: Creating a Sustainable Food Future, Installment Three', Aug 2013.

The proportion of the world population at below replacement level jumped from 46% to 68% in late 2021, when India's National Family Health Survey announced that the fertility rate has dropped to 2.1 in the countryside, where two-thirds of the population lives, and to 1.6 in cities and towns.

80 https://www.un.org/development/desa/en/news/population/world-population-prospects-2017.html

81 Kai Kornhuber, Dim Coumou, Elisabeth Vogel et al., 'Amplified Rossby waves enhance risk of concurrent heatwaves in major breadbasket regions', *Nature Climate Change*, 10, pages 48–53 (2020). https://doi.org/10.1038/s41558-019-0637-z

82 Michael E. Mann, Stefan Rahmstorf, Kai Kornhuber, Dim Coumou et al., 'Influence of Anthropogenic Climate Change on Planetary Wave Resonance and Extreme Weather Events', *Scientific Reports*, Vol. 7, Article number: 45242 (2017); K. Kornhuber, D Coumou, S Rahmstorf et al., 'Extreme weather events in early summer 2018 connected by a recurrent hemispheric wave-7 pattern', *Environmental Research Letters*, IOP Publishing 14 (2019) 054002; Kai Kornhuber, Talia Tamarin Brodsky, 'Future Changes in Northern Hemisphere Summer Weather

Persistence Linked to Projected Arctic Warming,' *Geophysical Research Letters*, 2021-02-28, DOI: 10.1029/2020GL091603

83   Paul R. Ehrlich, Carl Sagan and 19 others., 'The Long-Term Biological Consequences of Nuclear War,' Science 222, no. 4630 (December 23, 1983), pp. 1283–92; R.P. Turco et al., 'Nuclear Winter: Global Consequences of Multiple Nuclear Explosions', *Ibid.*; R.P. Turco et al., 'Climate and Smoke: An Appraisal of Nuclear Winter', *Science* 248, no. 4939 (January 12, 1990), pp. 166-76.

84   'India's Actions in Kashmir Risk Nuclear War', *Guardian*, 28 September 2019

85   Alan Robock, Owen B. Toon, Richard P. Turco and 7 others, 'How an India-Pakistan nuclear war could start—and have global consequences', *PNAS*, March 31, 2020 117 (13) 7071-7081; https://doi.org/10.1073/pnas.1919049117; Jonas Jägermeyr, Alan Robock and 17 others, 'A regional nuclear conflict would compromise global food security', *Ibid.* http://www.pnas.org/cgi/doi/10.1073/pnas.1919049117

86   https://www.mckinsey.com/featured-insights/future-of-asia/future-of-asia-podcasts/cop26-and-implications-for-asia

87   Statistics about 'terawatt-hours' in this paragraph and the next are from the Global Fossil Fuel Consumption chart in Our World in https://ourworldindata.org/fossil-fuels

88   Eric Holthaus, 'We've emitted more CO_2 in the past 30 years than in all of history.' *The Correspondent*, 16 October 2020.

89   Jonathan Watts, 'Ex-Maldives president to tell Cop26: do not compromise on 1.5C', *Guardian*, 1 November 2021

90   Paul Crutzen, 'Albedo Enhancement by Stratospheric Sulfur Injections: A Contribution to Resolve a Policy Dilemma?', *Climatic Change*, Vol. 77, pp. 211–220 (2006)

91   Andy Parker and Peter J. Irvine, 'The Risk of Termination Shock From Solar Geoengineering', *Earth's Future*, Vol 6, No. 3, pp. 456-467, March 2018 https://doi.org/10.1002/2017EF000735

92   Constantin W. Arnscheidt and Daniel H. Rothman, 'Routes to global glaciation', *Proceedings of the Royal Society A: Mathematical, Physical and Engineering Sciences*, Vol. 476 No. 2239 29 July 2020. https://doi.org/10.1098/rspa.2020.0303

93   Irvine, P. J., & Keith, D. W. (2020), 'Halving warming with stratospheric aerosol geoengineering moderates policy-relevant climate hazards,' *Environmental Research Letters*, 15 (4), 044011 (2020) doi:10.1088/1748-9326/ab76de Most of the research articles that led to this conclusion are cited in the above article.

94   Douglas G. MacMartin and Ben Kravitz, 'The Engineering of Climate Engineering', *Annual Review of Control, Robotics, and Autonomous Systems*, Vol. 2, pp. 445-467, May 2019, http://dx.doi.org/10.1146/annurev-control-053018-023725. See also Yan Zhang, Douglas G. MacMartin, Daniele Visioni, and Ben Kravitz, 'How large is the design space for stratospheric aerosol geoengineering?', *Earth System Dynamics*, Vol. 13, 201–217, 2022, doi.org/10.5194/esd-13-201-2022

95   Jacob T. Seeley, Nicholas J. Lutsko and David W. Keith, 'Designing a Radiative Antidote to CO_2', *Geophysical Research Letters*, Vol. 48, No.1, 16 January 2021. doi.org/10.1029/2020GL090876

96   Ru-Shan Gao, Karen H Rosenlof, Bernd Kärcher, Simone Tilmes, Owen B Toon, Christopher Maloney and Pengfei Yu, 'Toward practical stratospheric aerosol albedo modification: Solar-powered lofting', *Science Advances*, May 2021, 7 (20). doi: 10.1126/sciadv.abe3416.

97   Margaret Osborne, 'Here's How Wildfires Can Destroy the Ozone Layer', *Smithsonian Magazine*, 15 March 2023

98   UN Climate Press Release 25 Oct. 2021: Updated NDC Synthesis Report: Worrying Trends Confirmed. https://unfccc.int/news/updated-ndc-synthesis-report-worrying-trends-confirmed

Index

Grand Ethiopian Renaissance Dam (GERD) 215
Greenland 32–35, 181–4, 188, 271, 306
greenwashing 45
Gulf Stream (AMOC) 32–35, 82, 306

H

Hadley cells 217
Hansen, James 73–77, 82–84, 319
Harrison, Cheryl 160
Haywood, James 87, 292, 319
Hollub, Vicki 158
Horton, Josh 287–8
Hothouse Earth vii, 17–18, 25, 27–28, 30, 55–59, 61, 231, 310
Hunt, Hugh 32, 43, 320
hydroelectric power 102–3
hydrogen 29, 45–46, 70, 89, 93, 100, 105–110, 113–8, 126, 135, 166
hysteresis 35

I

ice 31–35, 51, 54, 115, 126, 169, 176, 180–192, 255, 262, 271, 291, 304
India 48, 99, 103, 129, 142, 153, 156, 198, 202, 205, 207–208, 214, 216, 221–9, 232–3, 254
Intertropical Convergence Zone (ITCZ) 101, 253
IPCC 5, 8, 27, 57, 74, 79, 83, 121, 151, 152, 158, 176, 190, 298, 300
Irvine, Peter 271, 276

J

Johnson, Matthew 144–6

K

Kashmir 216, 221, 333
Keefer, Bowie 185–6, 320
Keith, David 115–6, 248, 257, 271, 277–8, 287, 289–290, 320
Keller, David 163
Keskinen, Anton 264, 321
King, David 154
Kjellsson, Joakim 176, 177, 180, 321
Kornhuber, Kai 218, 321
Kravitz, Ben 272–5, 321

ozone 21, 70, 110–1, 240, 244–5, 281–2, 296, 308–9